IS SCIENCE ENOUGH?

IS SCIENCE ENOUGH?

FORTY CRITICAL QUESTIONS ABOUT CLIMATE JUSTICE

AVIVA CHOMSKY

BEACON PRESS
BOSTON

Beacon Press
Boston, Massachusetts
www.beacon.org

Beacon Press books
are published under the auspices of
the Unitarian Universalist Association of Congregations.

Printed in the United States of America

25 24 23 22 8 7 6 5 4 3 2 1

This book is printed on acid-free paper that meets the uncoated paper
ANSI/NISO specifications for permanence as revised in 1992.

Text design and composition by Kim Arney

Library of Congress Cataloging in Publication Data is available for this title.

To the next generation,
Ernesto, Mikel and Elio.
May we leave you
a better world.

CONTENTS

INTRODUCTION

We are facing a climate catastrophe. There are plenty of books, scientific reports, and articles out there that describe the damage we've already done—the droughts, the wildfires, the superstorms, the melting glaciers, the heat waves, and the displaced people fleeing lands that are becoming uninhabitable. Despite innumerable studies, proclamations, meetings, international agreements, and promises, despite scientific and technological advances in energy efficiency and alternatives, and despite huge investments in and the expansion of solar and wind energy and hybrid and electric vehicles, we denizens of planet Earth collectively continue to emit more carbon dioxide (CO_2) and other greenhouse gases into the atmosphere each year than the one before. The discourse might be going in the right direction, but our actions are going in, decidedly, the wrong direction.

The climate crisis is the result of a social and economic system that relies on extracting and consuming the earth's resources in ever-increasing quantities, and turning them into waste. Of course, every living creature consumes or absorbs resources and emits waste. But only humans have found ways to intensify the way we do it and developed ideologies and economies that promote unending expansion of the processes. Over the centuries, especially since the Industrial Revolution and the discovery of fossil fuels, we've made extraordinary leaps in production and consumption, as the world's population has grown.

But not all of us humans play the same role in this process. Control of fossil fuels allowed some groups of people to expand their power and standard of living. Over time, some of those who were excluded, dispossessed, enslaved, and exploited by the process fought for, and achieved, a piece of the growing political and economic pie. In the process, they also increased their own fossil fuel use and their emissions. Still, today, the process of expansion, dispossession, and incorporation continues. The masters of the system consume more than previous generations could have dreamed of. The growing middle classes consume significantly less, but still far more than their grandparents or the large population that lives in extreme poverty.

No matter how much our technological wizardry advances, the resources of the planet are finite, as is its ability to absorb the waste we produce. For the first couple hundred years, industrializing areas of the world managed to outsource a large chunk of the social and environmental costs of their production and consumption. They colonized distant lands, dispossessed and enslaved workers, and devoted a significant portion of their technological advances to instruments of war to ensure their continued dominance. And they displaced the consequences of their profligacy onto future generations.

By the late twentieth century, resources were growing scarcer, waste was becoming more toxic, and the planet's finite limits were finally being recognized. Scientists began to sound the alarm on the greenhouse effect, by which CO_2 and other gases released into the atmosphere by burning fossil fuels and by cutting down forests was trapping more of the sun's heat and warming the earth's temperature.

This book's contribution

Literature on climate change has proliferated over the past decade, as have popular awareness, media attention, political mobilization, and high school and university classes. Yet there is no single, short, accessible book that breaks down the complexities, terminology, disagreements, and issues in the debates for activists, students, and the general interested public. This book proposes to be such a primer.

I am not aiming at readers who question the science of climate change. Indeed, I will not delve into the science as there is already an ample literature that does this. Virtually every book on climate change, whatever its focus, begins with a chapter outlining what we know about the physical causes of climate change, the decades of science behind our knowledge, and the current and predictable future effects of increasing, stabilizing, or slowly decreasing CO_2 emissions.

Yet many questions remain largely opaque to general readers who understand we are facing a climate emergency but may be fuzzy on technical, policy, and social justice aspects. What kinds of policies could lower emissions enough to avoid impending climate chaos? What steps amount to little more than greenwashing? What issues divide the labor movement and the environmental movement, with respect to climate solutions? How does climate change relate to social and racial justice, including global and domestic economic inequality, poverty, migration, violence, and economic development? We may agree on the immediate material causes of global warming—greenhouse gas emissions. But what are the structural factors that keep us on what most agree is a path towards self-destruction? Where should we focus our attention to change the path?

Understanding climate change in order to change it

What's the cause of climate change? While the answer might seem obvious, this book argues that how we think about the causes of climate change matters a lot, because it shapes how we think about and organize to reverse an approaching climate catastrophe.

Global warming is caused by the increase in greenhouse gas (GHG) emissions that trap the sun's heat. The more we fill the atmosphere with these gases, the more heat is trapped. That's the technical explanation.

GHG emissions come primarily from the burning of fossil fuels, beginning with the extraction and burning of coal as the Industrial Revolution progressed in the nineteenth century and increasing through the extraction and burning of petroleum and natural gas in the twentieth. The late twentieth century saw massive increases in extraction of resources, and production and consumption of goods (for some), thus massive increases in the use of fossil fuels, bringing us to our current climate emergency.

If the science is now crystal clear that human activity—extraction, production, and consumption—causes GHG emissions that are progressively warming our planet and leading us towards catastrophic climate change, tracing this brief history suggests that the technical answer is only part of the picture. Growing use of fossil fuels may be the proximate "cause" of climate change, but what explains the growing use of fossil fuels? Is it population growth, which is bringing us close to eight billion inhabitants and growing fast, putting an ever-increasing strain on our planet's resources? Is it the inhabitants of the United States, who collectively burn about a quarter of the planet's fossil fuels every year, despite being home to less than 5 percent of its population? Is it the global 1 percent, the elite who overwhelmingly dominate high-emissions activities like air travel and luxury consumption? Or the top 10 percent, who produce over half of the planet's emissions, while the poorest half of the planet's inhabitants together produce less than 10 percent?[1] Is it the fossil fuel companies, which have worked so hard to deny the science behind global warming? Is it the governments that enact policies and agreements that continue to promote high-emissions activities? Is it the corporations that dominate the global economy and exert outsized control over governments and international agencies? Is it humans? Industrialization? Capitalism? How we identify "cause" will play a big role in how we conceptualize what it is that we need to change.

Climate change and planetary boundaries

While the impending climate emergency has captured a lot of attention, some argue looking at GHG emissions in isolation prevents us from seeing the bigger picture of the ecological crisis we are courting. Humans are exceeding planetary limits on multiple fronts, and the impacts of human activity in different areas are interrelated and create synergistic effects.

In 2009, the Stockholm Resilience Centre (SRC) proposed the concept of planetary boundaries—the outer limits of what humans can extract from or impose on the natural world in nine areas, each of which affects the others. "Crossing these boundaries," SRC argued, "increases the risk of generating large-scale abrupt or irreversible environmental changes."[2]

Climate change, measured by the amount of CO_2 equivalent in the atmosphere, is one of the areas in which the boundary is threatened. But it can't be separated from the others: ocean acidification, stratospheric ozone depletion, biogeochemical nitrogen (nitrogen released by agriculture and industrial processes), phosphorous cycle (phosphorous released into the oceans), global freshwater use, land system use (lands deforested and put into agricultural and urban use), biological diversity loss (species extinctions), chemical pollution, and atmospheric aerosol loading (air pollution, such as particles of dust and soot in the air we breathe). Exceeding any one of these boundaries "may be deleterious or even catastrophic," triggering "non-linear, abrupt environmental change within continental- to planetary-scale systems," the SRC concluded. When the center proposed the boundary system in 2009, it concluded we had already transgressed three of these boundaries: climate change, biodiversity loss, and changes to the global nitrogen cycle.[3]

The concept of planetary boundaries shows that despite the urgency of reducing GHG emissions to avoid catastrophic warming, it's not enough. If we succeed in reducing emissions but continue to pollute our waters, destroy our forests, and drive species to extinction, environmental disasters will persist. New infectious diseases like the 2019 novel coronavirus, on the heels of SARS, MERS, Zika, H1N1 ("swine flu"), Ebola, and others, are one example, resulting in large part from land use change as humans destroy wildlife habitats by pressing ever further into the planet's remaining forests.[4]

Organization of the book

The book divides its questions into five basic areas. The first section, on technical issues, looks at proposed and existing technical and technological responses to climate change. While existing and emerging technologies have brought indisputable benefits to many, technological advances have also brought social and environmental problems. Technology alone can't solve the problems we're facing, if it's incorporated in a social and economic system that prioritizes the profits of the few over the well-being of the many.

The second section explores the nature and implications of different policy options and proposals. Despite decades of international meetings and agreements, and despite a plethora of policy innovations, global emissions continue to rise. Why is this the case, and what kind of policies could be more effective? This section also looks at some of the economic and political interests that influence the policy discussions.

The third section asks the inevitable question in every conversation about climate change: "What can I do as an individual?" The specific questions here look at individual, consumer-based actions and different forms of protest and campaigns for change. While generally arguing lifestyle changes are a weak form of political activism, this section does not advocate one specific route towards bringing about change. Rather, it helps readers understand the debates surrounding specific types of activism and emphasizes the need for collaboration and complementary campaigns.

The fourth section focuses on social, racial, and economic justice, placing climate change in the context of global economic structures. Overall the world's poorest, who have contributed the least to global emissions, are suffering the greatest effects of the changing climate. This chapter explores divisions among social justice organizations, the labor movement, and environmental groups and ways these divisions could be transcended. It follows discussions in international policy circles and local organizations about how different countries, industries, and social sectors have contributed to climate change and asks how climate policy can fairly address these differences.

The fifth and final section delves into some of the biggest questions that the climate debate frequently evades. How are population growth and immigration related to climate change? Is capitalism inherently dependent on fossil fuels? On economic growth? Can there be economic growth without environmental destruction? What do the answers mean about what we need to change? How could we reorganize our world system to liberate people from poverty and hunger—and from the deadly treadmill of constantly increasing production and consumption? And, after decades of international, national, and local activism, are we making any progress?

Confronting climate change means understanding how we got to this point, and challenging some of the basic ways our society and economy are organized. The same systems and structures that have brought us to the environmental brink have forged our unstable and unequal world. I hope this book will give the interested public tools to more confidently and effectively engage in climate change debate and activism, including finding new forms of collaboration with movements for global justice.

CHAPTER 1

TECHNICAL QUESTIONS

CAN TECHNOLOGY SOLVE CLIMATE CHANGE?

This is a big question and is at the heart of many debates about climate change today. Techno-optimists, techno-utopians, or eco-modernists believe that even though modern industrial society is the cause of climate change, it can also be the solution. New technologies can solve the very problems we've created. It's just a matter of coming up with the right ones.

Most world governments, and businesses, have put a lot of faith in technological solutions to climate change. They flood us with visions of solar panels, driverless electric cars, artificial intelligence, smart grids, and carbon capture and storage technologies, and believe that the market itself can carry us to this carbon-free future. Others advocate government subsidies, incentives, regulation, or a combination of measures to encourage greater use of these new technologies. But techno-optimists believe that our basic economic model and system of production and consumption work fine, and that through technological progress, they can adapt to emit less.

Here I join critics who take a more radical, ecological, and systemic view. While technology must play a role in how we confront climate change, it's not enough to just incorporate new technologies and new regulations into existing economic models that are based on the plunder of the planet's resources. The global economy relies on ever-increasing consumption and economic growth, and the idea that new technologies will allow economies and consumption to continue to grow. But the model entails and is based on the exploitation of both the many and the planet, in the interests of the few.

An ecological approach

Instead of evaluating new technologies solely on their technical merit, an ecological approach focuses on systems and interrelationships. The science of ecology studies the relationship of organisms (plants and animals) and their habitats or environments. We often associate ecology with environmentalism: ecology insists on the interrelatedness of aspects of the natural world and shows how human-induced change and technology can have unwanted ripple effects that we miss by looking at one factor in isolation.

Ecological approaches also go beyond science. Political ecology considers how systems of human social organization interact with the natural environment: the co-production of the social and natural worlds, the relationship of politics and nature, and how the power structures of society and humans' relationships with the natural world shape each other. Ecological economics seeks to expand the discipline of economics, situating the economy in our "finite biosphere." It highlights the dependence of human-organized economies on the natural world that traditional economists often take for granted.[1]

Ecological approaches don't reject technology. Rather, they insist technology be understood within a web of political, social, and economic relationships that include those among humans, in the natural environment, and between humans and nonhuman nature.

Ecological approaches are radical in that they go beyond the surface to the roots. If new technologies are just imported into existing social and economic systems, the structures themselves will persist. If existing structures are based on privileging profit over people, finding more effective ways of extracting resources from nature, and endlessly increasing production and consumption, then new technologies controlled by the same powerholders will continue to uphold existing structures. This is why the slogan "System change, not climate change" emphasizes a much more fundamental transformation change than just adding new technologies to our toxic mix.[2]

Potentials and limits of technology

I'm not arguing that technology has *no* place in our collective attempt to reduce atmospheric CO_2 and rein in climate change. If new (and existing) technologies can be incorporated into a larger project of cultural, social, and economic change that allows humanity to share our resources fairly and within planetary boundaries, then they can contribute to this new world. In other words, technical innovation and diffusion should be used to support those who have the least, rather than those who already use too much. That's the political part.

Even though it's not the whole story, it's important to understand the technical aspects of how we got into this mess and what kinds of technical solutions have been proposed to get us out of it. The rest of the entries in this section look at these technical issues. First, I introduce some of the terms and concepts that we need to understand climate change. I then examine several technological innovations, from an ecological rather than a techno-optimist perspective, to suggest that it's unrealistic to believe we can rely on technical fixes alone to solve the climate catastrophe.

WHAT ARE GREENHOUSE GASES?

When we talk about greenhouse gases, we usually think first of carbon dioxide. Other greenhouse gases include methane, nitrous oxide, ozone, and fluorinated gases. They're called greenhouse gases because they trap heat in the earth's atmosphere, causing what has long been known as the greenhouse effect, now more commonly referred to as global warming or climate change. Some analysts believe that we should stop using these neutral-sounding terms and call it climate emergency, climate chaos, or climate catastrophe.

CO_2 is emitted in the largest amounts, though some of the other greenhouse gases are much more potent than CO_2. Some statistical analyses only measure CO_2, while some use the concept of CO_2 equivalence (CO_2e)—that is, they translate the impact of the others into their CO_2 equivalent for simplicity's sake. If an analysis measures only CO_2, it's necessarily underestimating the actual impact of greenhouse gas (GHG) emissions.

The US Environmental Protection Agency (EPA) estimates that CO_2 accounts for just over 80 percent of US GHG emissions, methane for 10 percent, nitrous oxide for 6 percent, and fluorinated gases for 3 percent.[3] Much methane is released accidentally, and much of that is not measured, so many scientists believe methane emissions are significantly higher than the EPA figures.

CO_2 is released into the atmosphere primarily through the burning of fossil fuels—coal, natural gas, and petroleum—but also through burning renewable sources like wood and biomass, and through waste incineration, whether in waste-to-energy operations or simple incinerators. Plants also factor into the CO_2 cycle. As they grow, they absorb CO_2 through photosynthesis, removing it from the atmosphere. This is why forests are called a carbon sink. But when a forest is felled, the carbon stored in the plants is released into the atmosphere as they decompose.

Although methane is released into the atmosphere in smaller quantities than CO_2, it's much more potent in its impact on the climate. Methane doesn't remain in the atmosphere as long as CO_2, but in the first twenty years after it is released, its warming impact is eighty-four times that of CO_2. If measured over a hundred years, methane's impact is around thirty times greater. Scientists who measure atmospheric methane estimate it's responsible for about 15 to 25 percent of recent global warming.[4]

Methane emissions result from the production and transport of fossil fuels (39 percent, especially from natural gas), from livestock and other forms of agriculture (36 percent, especially from livestock digestive processes), and from solid waste decay in landfills (16 percent). Official emissions statistics seriously underestimate the methane lost to leakage and accidents as natural gas is extracted—increasingly through fracking—and transported around the country through a vast system of pipelines. Scientists are still debating exactly why atmospheric methane has risen so sharply since 2006. Some believe over one-third of the global increase is due to the growth of the fracking industry in the United States and Canada. Climate change may also increase the release of methane from freshwater systems and from thawing permafrost.[5]

Nitrous oxide is released in much smaller quantities than CO_2 or methane, but it's almost three hundred times as potent as CO_2. Although there are multiple sources, the largest by far is the production and use of agricultural fertilizers (73 percent).

Some of the most dangerous greenhouse gases of all are the fluorinated gases developed to replace ozone-depleting chlorofluorocarbons (CFCs), which were banned in the 1980s. Although released in much smaller quantities, fluorinated gases including CFCs, hydrofluorocarbons, hydrochlorofluorocarbons, perfluorocarbons, and sulfur hexafluoride have by far the highest potency of all the greenhouse gases—thousands or tens of thousands of times that of CO_2. And they can last in the atmosphere for thousands of years. Because of their extreme impact, they are sometimes referred to as high global warming potential (HGWP) gases.

WHAT ARE THE MAIN SOURCES OF GHG EMISSIONS?

The United States emits about 6.5 billion metric tons of CO_2 equivalence every year; the entire world produces about 50 billion metric tons a year.[6] More later on the disproportionate role the United States plays in global carbon emissions. For now, let's look at where those 6.5 billion metric tons come from.

Consumers are probably most aware of their use of fossil fuels when they fill up their gas tanks, buy heating oil, pay for natural gas, and purchase electricity, which in the United States most likely comes from a plant that burns coal or natural gas. But fossil fuels, and their inevitable emissions, are built into other aspects of our lives as well.

Power generation accounts for about a quarter of US emissions, just below the top emitter: transportation. These two are followed by industry, at 23 percent of total US emissions; commercial and residential buildings, at 13 percent; and agriculture, at 10 percent. The industry and buildings figures measure fuel use on-site. That is, they measure the oil or natural gas used in industrial processes, or for cooking, heating buildings, and warming hot water. If industry and building electricity use were counted toward those sectors' emissions (rather than towards the power generation sector), the two would count as the country's top emitters, at 30 and 31 percent.[7]

Of course, fossil fuels are the main culprit everywhere. But we need to think about eliminating fossil fuel use in all of these areas, not just in the plants that generate our electricity.

The grid

The electrical grid—power generation plants and transmission infrastructure—is important because it's a significant source of emissions, even if not the largest. And it's important because some of the other sectors' emissions could be reduced if they were electrified and could rely on a cleaned-up grid. Some localities are trying to reduce building emissions by banning gas- and oil-fueled appliances and heating systems.[8] Switching to electric could bring big reductions in emissions, if the electricity is produced by clean renewable sources. The same is true for electric cars: with a clean grid, powering cars by electricity instead of gasoline would significantly reduce emissions. But that's a big "if," given the current state and trends in the grid.

Cleaning up the grid by shifting from coal to gas

As of 2020, the main source of power generation in the United States was natural gas (40 percent), followed by coal, nuclear, and renewables (around 20 percent each).[9] Fossil fuels still account for 60 percent of the energy we produce.

Promises from President Donald Trump to bring back the coal industry notwithstanding, the main way the United States stabilized or lowered emissions during the Obama and Trump administrations was by promoting the natural gas industry, at the expense of coal. With aggressive investment in new fracking technologies, which started in the 2000s, supplies of natural gas rose and prices dropped. Coal use in power generation has been declining steadily since its peak in 2010, as natural gas has risen to take its place, bypassing coal as the major source in 2016. (Much smaller increases in the use of renewables have barely made a dent.)[10] To the extent that we've seen declines in GHG emissions in the United States, the shift from coal to natural gas in power generation is one of the main factors.

Some coal companies have struggled or gone bankrupt during the transition. The closing of coal mines has been devastating for the workers and communities that relied on them for economic survival. But the shift from coal to gas has overall been a gift to the fossil fuel industry. Some coal producers, like Exxon and Chevron, sold off their coal mines to devote new investments to gas. Other fossil fuel giants, including Shell and BP, and state-owned companies, like Russia's Gazprom and China's National Petroleum Company, have also thrived during the natural gas boom.[11]

Thus far, the grid has offered a kind of low-hanging fruit for those who hope that we can juggle power sources, moving to a cleaner fossil fuel while keeping the rest of the system and our lives intact. But replacing one fossil fuel with another in the generation of electricity is a stopgap measure. Natural gas has environmental costs too, and as its use rises, so will those costs. And the grid is only one piece of a much larger panorama of GHG-emitting sectors.

Beyond the grid: Transportation

While technical, policy, and economic factors have succeeded in bringing down emissions from the power grid somewhat, the same is not true for the largest US emitter, the transportation sector. A little over half of the transportation sector's emissions come from passenger vehicles, including cars, SUVs, light-duty trucks, pickup trucks, and minivans. Other sources include freight and other trucks (23 percent), commercial aircraft (9 percent), ships, boats, and trains.[12]

Attempts to reduce emissions from the transportation sector have focused primarily on increasing fuel efficiency, with mixed results. Between 1990 and 2004 average fuel economy actually *decreased*, as sales of SUVs outweighed improvements in passenger cars. After that, light truck sales leveled off, though they still accounted for 52 percent of new vehicles purchased in 2018. Meanwhile, passenger miles driven increased 46 percent between 1990 and 2018, offsetting any improvements in efficiency.[13]

Electric cars are being promoted in Europe and China to reduce transportation emissions, and the Biden administration began to bring the United States on board with toughened emission standards and an August 2021 executive order requiring that half of new cars sold be electric by 2030. After decades of resistance—and with the promise of significant federal incentives—the US auto industry has started to move creakingly in that direction. Some fear that investing further in individual automobility only distracts us from more structural approaches to reducing emissions in the transportation sector, like "sustainable urban environments, micromobility, and mass transit."[14]

Beyond the grid: Industry

Three key industries account for 58 percent of all industrial sector energy use in the United States: chemical, mining, and refining. These industries use electricity from the grid, and they also use fossil fuels to produce their own electricity and to run their boilers and industrial processes. Mining and refining industries burn fossil fuels to extract and process more fossil fuels. The chemical industry, in addition to burning them as a power source, uses fossil fuels as a raw material to produce things like plastics and fertilizers.[15]

Matthew T. Huber reminds us to consider what he calls the "vast and largely invisible middle ecologies of processing and industrial resource consumption" and critiques a "politics of climate" that "ignores the massive contributions of the industrial sector. If we were to focus our policy and political energy on transforming how the industrial sector consumes energy and emits carbon, we would tackle a massive chunk of the entire societal carbon footprint." The ammonia (fertilizer) plant in Louisiana that Huber studied is the nation's top chemical sector emitter, with about 60 percent of its emissions coming from combustion and 40 percent from the chemical process that turns natural gas into ammonia-based fertilizers, emitting CO_2 in the process.[16]

Cement—the main ingredient in concrete, the most widely used construction material in the world—is another major culprit, accounting for 8 percent of the world's CO_2 emissions in 2018. "If the cement industry were a country," the BBC reported, "it would be the third largest emitter in the world—behind China and the US. It contributes more CO_2 than aviation fuel (2.5 percent) and is not far behind the global agriculture business (12 percent)." Forty percent of the industry's emissions come from burning fossil fuels, 50 percent from the chemical process.[17]

Emissions from the chemical processes in these industries and others, like glass and chemical products, are usually not even measured in calculations of countries' CO_2 emissions. One study estimated unreported CO_2 emissions from China's production of five industrial products (alumina, plate glass, soda ash, ammonia, and calcium carbide) came to 233 million metric tons in 2013—equivalent to the entire (measured) emissions from the country of Spain.[18]

Beyond the grid: Agriculture

Agriculture contributes 10 percent of US CO_2 emissions, an amount less than some other sectors but still significant. And the agricultural-industrial complex is intertwined with the fossil fuel economy in numerous ways. If we apply a comprehensive life-cycle analysis, the food production sector jumps to some 20 to 40 percent of global GHG emissions.[19]

A life-cycle accounting would begin with deforestation and land use changes. It would include the production of farm machinery, fertilizers, and pesticides. It would go on to transportation and food processing and end with supermarkets. Every step of our food production and distribution system depends on fossil fuels.

Livestock raising accounts for over half of agricultural emissions worldwide. Livestock-related emissions include methane from the digestive processes of the seventy billion animals raised for human consumption, nitrous oxide produced by their manure, and CO_2 from managing the grazing and cropland for their feed.[20]

What we don't measure

Even if we broaden our scope beyond the grid to address emissions in transportation, industry, buildings, and agriculture, we're still missing part of the picture. There's a lot that's hidden behind and around the statistics in the emissions reports provided by the US and other world governments. Some sectors aren't officially counted at all, although scholars have tried to find ways to calculate what they contribute to emissions. International trade also complicates the picture. Policymakers argue about how to incorporate emissions from trade itself (i.e., shipping) and whether emissions should be counted towards the countries that produce the items for export, mainly manufactured goods and agricultural products, or towards the countries that import and consume the items.

Emissions that escape the measures: Military

One area not included in official statistics is the fossil fuel emissions caused by military activities around the world. In fact, the

United States insisted that military emissions not be included in reporting and reductions required by the Kyoto Protocol, which it didn't sign anyway. The Paris Agreement ended the automatic exemptions of the world's militaries from reporting, but still left it up to individual countries to determine any reductions. The US military's "carbon boot-print" dwarfs the emissions of many of the world's countries.[21]

The US Department of Defense emitted 34.5 million metric tons of CO_2 equivalence from "non-standard operations" (i.e., military operations) and 21 million metric tons from "standard operations" (e.g., maintaining facilities) in fiscal year 2018, far more than any other government agency. One study calculated emissions from the first five years of the Iraq War, from 2003 to 2008, at 141 million metric tons, noting that the war alone emitted more CO_2 per year than did 139 countries. None of this appears in official US emissions reports.[22]

If we measure all war-related and other Department of Defense fuel use, "the DOD is the world's largest institutional user of petroleum, and correspondingly, the single largest producer of greenhouse gases in the world." Between 2001 and 2017, the US war machine emitted 1,212 million metric tons of CO_2, about 766 million of which was in military operations, concludes Neta Crawford in a Brown University study. Much of this fuel, paradoxically, is dedicated to guaranteeing control of Persian Gulf oil.[23]

Jet fuel is by far the biggest source of military consumption, followed by facilities maintenance and diesel fuel. Five hundred US military installations that occupy 27 million acres of land at home and abroad and include 560,000 buildings account for about 30 to 40 percent of DOD emissions. To capture the actual extent of fuel use by the US war machine, argues Crawford, we need to include emissions caused by military-related industries (153 million metric tons a year) as well as infrastructure destruction and reconstruction, targeting of oil wells, and deforestation, which are extremely difficult to calculate.[24]

Emissions that escape the measures: Aviation

Also left out of international emissions reporting are the international aviation and shipping industries, in part because they are difficult to attribute to a single country, so they don't fit neatly into the system of national-level reporting. These industries are powerful interests in the global economy that policymakers are reluctant to challenge. Together, shipping and aviation account for about 5 percent of global emissions. Emissions from shipping were on track to rise between 50 and 250 percent by 2050 and aviation emissions to double or triple.[25]

If the global aviation sector were counted as a country, it would be the sixth-largest global emitter. Eighty-one percent of the world's flights are for passengers (as opposed to shipping freight), and two-thirds of those flights are domestic. Since domestic flights tend to be shorter than international flights, the latter still account for 60 percent of passenger flight emissions.[26]

Perhaps unsurprisingly, flight emissions patterns follow those of fossil fuel use. The largest culprit is flights originating in the United States, which accounted for 24 percent of all passenger flight emissions, followed by China, the UK, Japan, and Germany. "Less developed countries that contain half of the world's population accounted for only 10% of all emissions," one study points out. Even within the high-emitting countries, a small portion of the population accounted for the lion's share of flights. In the UK, 15 percent of the population took 70 percent of the flights. In the United States, the 12 percent of the population who took more than six flights a year accounted for two-thirds of all emissions, while half the population doesn't fly at all.[27]

Yet international climate accords and national policies have made no attempt to implement strategies to reduce air travel. Instead, the aviation industry and its organizations float general ideas about voluntary increases in fuel efficiency and alternative fuels, while promoting feel-good accommodations like carbon offsets.[28]

Can we reduce emissions in all these sectors?

Science has succeeded admirably in identifying the different sources of GHG emissions. But to figure out how to reduce emissions enough, and quickly enough, to slow the climate emergency, science is not enough. As a country and in the world as a whole, we have so far been attacking the problem of emissions in a piecemeal fashion, hoping that multiple technical fixes will reduce emissions in different sectors enough to avert catastrophic global warming. But given the magnitude and scope of the problem, these fixes have failed to alter our trajectory, even as time is running out.

HOW CAN WE CLEAN UP OUR ENERGY GRID? WHAT EXACTLY ARE "CLEAN," "RENEWABLE," AND "ZERO-EMISSION" ENERGY SOURCES?

Many proposals for addressing climate change have focused on the plants that produce our electricity. President Obama's 2015 Clean Power Plan required states to gradually reduce their CO_2 emissions from power generation but allowed them to determine how to accomplish the mandated reductions. Plants could still burn fossil fuels but might install upgrades to reduce emissions or switch from coal to natural gas. Legal challenges, and Trump administration initiatives, derailed the plan, but some states pushed ahead with their own comparable requirements. The 2019 Green New Deal House Resolution took a much stronger position, calling for "100 percent . . . clean, renewable, and zero-emissions" sources in power generation within ten years. President Joe Biden promised a "carbon pollution-free power sector by 2035."[29]

You might think we're talking about the same thing when we refer to mandating reduced emissions and to "clean," "renewable," and "zero-emission" energy sources. But they're not necessarily the same thing.

Fossil fuels, which currently supply about two-thirds of US energy usage, are definitely not renewable or zero-emission. They're nonrenewable because they were formed over millions of years and exist in a finite supply, so we are using them up. They're high-emission because when they are extracted, transported, and burned, they emit carbon dioxide, methane, nitrous oxide, and other greenhouse gases. They are also dirty because of the smoke, fumes, and particulates that are released when they are burned. Still, many environmental organizations consider natural gas to be "clean" because it produces less local pollution and lower CO_2 emissions than coal.

Renewables, on the other hand, include sources that can be used without being used up, like the sun and the wind, and sources that can be replenished, like trees, biomass, and agrofuels or biofuels. Other sources, like hydropower, are a bit controversial; while the EPA considers it a renewable source, some states and environmental organizations don't. To access the power of naturally running water, rivers are usually dammed to create storage reservoirs, from which water is pumped as needed. As long as the sources of the river remain, water should continue to flow. But the mammoth dams and reservoirs change water flows, affecting fish and other wildlife, water quality, and water availability. And the changing climate itself may shrink the water sources that hydropower relies on.

Not all renewables are clean or zero-emission when they are used to produce electricity. Biomass and agrofuels are renewable in that new plants can grow to replace those burned. But they still create CO_2 emissions and other pollutants when they are burned to produce electricity. And since the amount of cultivable land on the planet is finite, the plantations where agrofuels are sourced entail deforestation and other land use changes—making them less than fully renewable.

Solar and wind power might seem to fulfill all three requirements. They are renewable, they are clean, and they produce electricity without creating emissions. However, while the *sources* of these kinds of energy may be renewable and the power-generation process is clean and zero-emission, the infrastructure for harnessing and transforming sunshine and wind into usable energy and storing and distributing that energy are not necessarily so. These processes require extraction of resources, changes in land use, and creation of waste, all of which can produce GHG emissions and other sorts of environmental harms.

When analysts measure emissions from different energy sources, they are usually referring to emissions during the process of producing electricity. But a better calculus is based on life-cycle emissions. Life-cycle calculations look at emissions beginning with the extraction of raw materials and including the building of processing, generation, storage, and transmission systems, and eventual disposal of waste. Life-cycle calculations make it clear that no form of energy production is really emissions-free—though there are huge differences among the various forms.

The scale of the dams and storage needed for major hydropower projects means that these projects often require flooding residential, farm, and forested lands, and reducing water flow to areas downstream. Human, fish, and other wildlife habitats are destroyed. The world's largest hydropower project, the Three Gorges Dam in China, displaced over a million people and destroyed hundreds of towns and villages. In Southeast Asia, a wealthy few are benefiting from a hydropower boom as a series of dams reduces the Mekong River to a trickle and "a lifeline for 60 million people is being choked." Some three hundred thousand farmers and fishers are displaced from its delta every year as their source of subsistence disappears. Hydropower's man-made reservoirs also generate greenhouse gases of their own, in particular methane and nitrous oxide, through decomposing organic matter. Yet these emissions are not generally included in countries' accounting, making hydropower appear more benign than it actually is.[30]

When the New England Clean Energy Connect project proposed to build transmission lines through Maine to import hydropower from Quebec to the New England grid, many local and national environmental groups opposed the project. Its "clean" name, they argued, was little more than greenwashing—a way to paint a green veneer, at little cost, on business-as-usual. As one study concluded, "While large scale Canadian hydroelectricity may be renewable, it's not green."[31]

The proposed transmission line would severely impact forested areas of northern Maine. And while it would allow New England consumers to purchase their power from a renewable source, it wouldn't necessarily reduce fossil fuel use overall, because unless production was greatly expanded, Hydro-Quebec's existing customers would need to shift to fossil fuels as soon as their hydropower was diverted to New England.[32]

Increasing hydroelectricity production in northern Quebec would also have environmental and social costs. For decades, Indigenous peoples in the region have been protesting against dams that have destroyed their territories for the profits of Hydro-Quebec and to burnish the green credentials of New England and New York electricity consumers. When New York mayor Bill de Blasio traveled to northern Quebec in 2019 to meet with Indigenous leaders about the impact of a potential deal to purchase Quebec hydropower for New York City, Bill Namagoose of the Cree Nation government pointed out that it was the first time a purchaser had ever investigated the impact of Hydro-Quebec's works. "The Crees have said they don't want any more Hydro projects on their territory, and that's the message we gave to the guests from New York," Namagoose explained.[33]

Unlike hydropower, producing energy from the sun and the wind, through solar panels or windmills, doesn't affect the availability of the resources themselves. But they too create emissions and have other environmental and human impacts beyond the moment of electricity generation. They require land and infrastructure like solar panels, windmills, and storage and transmission systems that are produced in factories, with resources extracted from the earth.

It's true that solar panels don't emit any greenhouse gases or other contaminants as they generate electricity. A life-cycle calculation, though, includes the emissions involved in the production, transport, and eventual disposal of solar panels, which last about forty years. Divided over the life span of the panel, solar panels' CO_2-equivalent emissions are estimated at approximately 50 grams of CO_2 equivalent per kilowatt-hour of electricity. Compare this to natural gas (500 to 600 grams of CO_2 equivalent per kilowatt-hour) or coal (1,000 to 1,200 grams of CO_2 equivalent per kilowatt-hour).[34] Even with a life-cycle accounting, solar power produces emissions that are quite low, though not exactly zero.

Despite the low emissions of solar panels, other environmental costs involved in their production and disposal make them less than perfectly "clean" and not completely "renewable." Like all electronic devices, solar panels require the mining of toxic materials, create pollution in their production, and end up as toxic e-waste. If the price of solar panels has fallen, it's partly because more of them are being produced in China and other low-wage, low-regulation countries, where human and environmental costs are magnified and the quality and life span of the panels may be lower. And as we scale up the use of solar panels, the more waste we create. Solar farms also change landscapes and require storage and transmission infrastructure that comes with its own environmental costs.

Most critiques of the environmental costs of solar power come from sources seeking to promote fossil fuels or nuclear energy. While their larger argument makes little sense—solar's impact is far less harmful than that of those two alternatives—the fact remains that solar does use finite resources and create waste.[35]

Like the sun, the wind can be harnessed to produce electricity with no carbon emissions. And as with solar, emissions and other impacts are produced over the life cycle of wind energy production. Land has to be cleared for wind farms, raw materials have to be extracted, windmills have to be built, energy has to be stored and moved to where it's needed, and eventually the windmills have to be replaced. But life-cycle emissions from wind energy is even lower than for solar—between 10 to 20 grams of carbon dioxide equivalent per kilowatt-hour.[36]

Yet wind farms have provoked protest on environmental grounds because they occupy significant space and affect the on- or off-shore environments where they are sited, including those of fish and other wildlife, and because of the noise pollution they create. In many ways, wind power is the most environmentally benign form of power generation. However, estimates vary widely as to how much land needs to be dedicated to wind farms in order to have enough of them to fully supply US power needs.[37]

Wind and solar share the additional characteristic of being intermittent; panels or turbines produce energy only when the wind is blowing or the sun is shining. Both, then, require storage systems—basically, giant batteries—in order to keep energy consistently available to consumers. Thus, they push us into what political scientist Thea Riofrancos calls "the extractive frontiers of the renewable energy transition" in the search for lithium, cobalt, manganese, and graphite for battery production.[38]

If we intend to maintain our current levels of electricity use with sun and wind power, we're talking about creating a new set of environmental disasters in the mining of raw materials for batteries and in the disposal of used batteries. "Going 100% renewable power," reported *Grist*, "means a lot of dirty mining." South American lithium mines suck up huge quantities of water, displace farming communities, and release lung-damaging dust. In Chile's Atacama Desert, which holds some 30 percent of the world's supply of lithium, local organizations are protesting "'green extractivism': the subordination of human rights and ecosystems to endless extraction in the name of 'solving' climate change." Meanwhile, most of the world's cobalt comes from the Democratic Republic of Congo, where forty thousand children work in perilous conditions. Mines spew toxic dust and waste, including uranium. Graphite mines also contaminate and displace local communities. Other studies have detailed the hidden costs of disposing of these batteries.[39]

Finally, agrofuels or biofuels offer a source of energy that is, in theory, renewable but is not necessarily "clean." Agrofuels create multiple, complex impacts on carbon emissions. I say that they are "in theory, renewable" because their cultivation requires land, as well as inputs like farm machinery, pesticides, and fertilizers. This means either that land currently devoted to growing food must be diverted to growing crops for agrofuels or that forests must be felled to bring new lands into cultivation. After visiting Borneo, Indonesia, and witnessing the clear-cutting of forests and village lands for massive palm plantations to produce oil for biofuel, journalist Heather Rogers called biofuels "among the most disastrous solutions out there." "Precisely when we need forests to stay intact and perform the vital function of absorbing and storing CO_2, they're being ripped away in the name of stopping global warming," she concludes.[40]

Biofuels can be used for electricity generation, and the federal Renewable Fuel Standard program mandates that plant-based fuel be mixed into the US gasoline supply—primarily corn ethanol and soybean biodiesel. These requirements have led to significant expansion of corn and soybean production. Both are industrially produced crops, heavily reliant on irrigation, fertilizers, pesticides, and machinery—all of which create CO_2 emissions and other environmental impacts. Pollutants and emissions are also released in the ethanol refining process.[41]

While energy production from plant sources is not "zero-emission" either—they release greenhouse gases when they are burned—it's sometimes considered a "carbon-neutral" process because plants absorb CO_2 over their life span. But the balance varies considerably, depending on farming techniques and prior land use, as well as on refining technology, transportation, and so on.

As researcher Raj Patel explains in his foreword to *Eating Tomorrow* by Timothy Wise, "There is a difference between renewable resources and sustainable ones. Sure, the corn will come out of the ground this year, and then come out of it next year. It is renewed. But the system into which corn-based ethanol plays, mortgaged to the internal combustion engine and the fossil fuel industry" is anything but sustainable. Wise concludes that first-world demand for biofuels has increased food prices and hunger among those who can no longer afford the foods they used to grow themselves or purchase at low cost.[42]

Probably the most controversial zero-emission energy source is nuclear power. Because nuclear power generators do not emit any greenhouse gases, some environmentalists advocate expansion of nuclear power as an alternative to fossil fuels. But environmental organizations also have a long history of opposition to nuclear power because of its radioactive waste, the threat of nuclear meltdown, such as occurred in Chernobyl and Fukushima, and the dangers involved in mining uranium. Emissions over the life cycle of nuclear power production occur up and down the line, from mining to transport and waste disposal. Yet nuclear energy is dubbed "clean" or "zero-emission" (though not "renewable") because in its daily operations a nuclear plant emits neither greenhouse gases nor other local pollutants.

Every energy source has its costs. We need to accept that there is really no source of energy that is truly clean, renewable, or zero-emission, much less any that fulfills all three goals. Instead, we should speak more precisely—about energy sources that are cleaner, more renewable, or lower-emission. No matter what the energy source, the more we increase its use, the greater its emissions and other environmental impacts will be.

The belief that we can solve our emissions problem by developing cleaner forms of energy sidesteps another issue. Under current political and economic arrangements, there is no evidence that increasing the use of cleaner energy sources decreases our use of fossil fuels or lowers our emissions.

Right now, we inhabitants of planet Earth are collectively and steadily increasing our use of energy. Thus, even as we've added more alternatives to the mix, they haven't displaced fossil fuel use at all. Instead, we're using more alternatives *and* more fossil fuels. It's the worst of both worlds: fossil fuel use goes up, and use of other sources, with their different costs, is also increasing.

If we want to keep the planet habitable for future generations, we need to think a lot bigger. As we develop energy sources that are cleaner, lower-emission, and more renewable than fossil fuels, we need to work on ways to also significantly reduce our use of energy, not just add new sources. That means political and economic change.

The burning of fossil fuels is causing dire and immediate dangers to the planet. While every energy source has social and environmental impacts—some of them potentially devastating—drastic and immediate reduction in the use of fossil fuels is essential. But looking for new sources of energy isn't going to get us there if we keep increasing our use of energy.

Meanwhile, almost a billion of our planet's people have no access to electricity at all, and three billion lack access to cooking fuel.[43] Any energy transition must also address the issue of access. We need to transition (to cleaner sources), reduce (overall), *and* redistribute for a truly sustainable global energy system.

WHAT'S THE DIFFERENCE BETWEEN "ZERO-" AND "NET-ZERO" EMISSIONS?

While they sound almost the same, there's a big difference between "zero" and "net-zero." "Net-zero" means that greenhouse gases can still be emitted, if they can be "offset" by other activities that remove CO_2 from the atmosphere, like increasing forest cover or other, more controversial high-tech proposals to capture and store carbon.

The concept of "net zero" makes sense because it's inevitable that there will be *some* GHG emissions. As we've confirmed, calling any energy source "zero emissions" is not really accurate. Even if there are few or no emissions from the operation of the system—as with solar or wind energy—there are emissions across the life cycle. Nor can industry, agriculture, transportation, or other sectors become completely emission-free. Offsetting inevitable emissions may be a useful goal.

Offsets get problematic, though, when they become a rationale for continuing business as usual. "The idea of net zero has licensed a recklessly cavalier 'burn now, pay later' approach which has seen carbon emissions continue to soar," wrote three climate scientists in 2021 to explain why they could no longer advocate the net-zero approach. Schemes that propose to remove carbon from the atmosphere ranged from "fantasy" to "pipe dreams" and have encouraged scientists and politicians to avoid confronting the real problem: global emissions must go down, immediately and dramatically, if we want to slow the already-catastrophic impacts of the warming climate. Our priority should be on reducing emissions from all sectors as much as possible, rather than using net-zero abstractions to justify ever-growing emissions.[44]

CAN FORESTS SERVE AS A CARBON SINK?

One low-technology, proven way to remove CO_2 from the atmosphere is through the natural process of CO_2 absorption by plants. The world's forests currently absorb about 30 percent of annual carbon emissions.[45]

Plants nourish themselves through photosynthesis—using sunlight to convert water and CO_2 into chemical energy that sustains life and growth. In the process, they absorb CO_2 from the atmosphere and store or sequester it. Large plants like trees, and large concentrations of plants like forests, absorb and store more CO_2. Trees store CO_2 in their leaves, branches, trunks, and roots, and they continue to absorb CO_2 as they live and grow. Forests also store CO_2 in their soil. When forests burn or are cut down, much of this stored CO_2 is released into the atmosphere. And forest loss means that trees are no longer absorbing atmospheric CO_2 and no longer helping to regulate local temperatures by providing shade and transpiring water.

The world lost 420 million hectares of forest between 1990 and 2020.[46] Net loss—when the area reforested is subtracted from the total hectares lost—was about 178 million hectares, and reforestation rates have been increasing. But newly reforested areas don't absorb as much CO_2 as the older forests that were destroyed, a fact obscured by these statistics. And some of what's counted as reforestation are tree plantations that will soon be felled for lumber.[47]

All forests absorb CO_2, but tropical forests are especially important and especially at risk. Because of their rapid growth, tropical forests absorb more CO_2 than do slower-growing forests in colder climates. Because they are almost all located in the Third World, they may be more at risk because of development imperatives and lack of regulation.

Fire, logging, expansion of large-scale agriculture, ranching, and mining all contribute to deforestation. Other causes of forest loss include drought, hurricanes, and insect infestation, which are sometimes called "natural" phenomena but are also exacerbated by human-caused climate change.[48]

One 2017 study determined tropical forests are now actually emitting more carbon than they are absorbing, due to ongoing forest destruction. "If tropical deforestation were a country," the World Resources Institute concludes, "it would rank third in carbon-dioxide equivalent emissions, only behind China and the United States of America." (This figure, though, measures only the greenhouse gases released by deforestation, not the CO_2 absorbed by remaining forest.) Another study found emissions from deforestation essentially balance out ongoing sequestration by remaining tropical forests, leaving their overall CO_2 impact as neutral as of 2019.[49] In 2021, the Amazon rainforest crossed the threshold and began to emit more CO_2 than it was absorbing.[50]

Some critics question the emphasis on tropical forests: it reflects a colonialist mentality that Third World countries aren't capable of managing their resources and that Third World resources should be managed in the interests of First World corporations and consumers. Industrialized countries cut down their forests for their own economic development, now they want to "offset" their emissions by claiming control of Third World forests.

It's true that Global North governments, corporations, and organizations currently use Third World projects as a form of greenwashing, attempting to divert attention from their own environmental impacts. In fact, their involvement in Third World deforestation runs very deep. First World industrialization was an explicitly colonial process, which relied on Third World deforestation from its earliest days. (Think Caribbean and Brazilian sugar plantations, African minerals, timber, and ivory, Malaysian rubber, and so on.) Furthermore, much current Third World deforestation is carried out by multinationals producing cattle, soy beans, and palm oil for export—in part to help industrialized countries reduce their own carbon footprint. Third World Indigenous and peasant organizations are fighting *against* the deforestation and extractive projects carried out by local elites and foreign multinationals, even if their governments are sometimes willing partners in these neocolonial endeavors.

To avert climate catastrophe, we need to both reduce emissions and find ways of sequestering emitted CO_2. Forest preservation, afforestation (planting new forests) and reforestation require no major advances in technology: we already have the knowledge and the tools to pursue them. But governments, agencies, and organizations rarely prioritize forest protection in their climate change mitigation plans. Only 3 percent of global climate mitigation funding goes to tropical forest preservation, even less goes to northern forests. Emissions from deforestation, like emissions from other sources, have risen since the Paris Agreement was signed at the end of 2015. While forests overall still serve as a "sink," the size of the sink is shrinking.[51]

WHAT IS CARBON CAPTURE? IS IT A VIABLE SOLUTION?

Many techno-optimists are less interested in low-tech options like forest preservation and more interested in new technologies that will, they hope, be able to "capture" CO_2 as it's emitted from power plants and other sources, and thus prevent it from entering the atmosphere. The fossil fuel industry is a major proponent of carbon capture, since it's a way to appear climate-friendly while continuing to burn fossil fuels.

What happens to the CO_2 after it's captured? Proponents have proposed two potential options—carbon capture and storage (CCS) and carbon capture, utilization, and storage (CCUS). With CCS, the CO_2 would be transported through pipelines and injected and stored deep underground in exhausted oil and gas fields or coal mines, or in the ocean or seafloor. Or it could be fixated into inorganic carbonates like limestone. CCUS proposes commercializing captured CO_2. Paradoxically, its one currently viable use is in enhanced oil recovery—a process similar to natural gas fracking in that it injects chemical substances deep into the ground to access oil deposits otherwise inaccessible. So CO_2 can be captured and used to create even more emissions. Or it could theoretically be incorporated into the manufacture of concrete or other products. Or used as a source of feed for farmed fish or in the production of biomass.[52]

Most of the technology for CCS or CCUS is embryonic at this point. The European Union reported discouraging results on $500 million designated towards developing and implementing CCS technologies as part of its post-2008 recovery package. Despite the investment, CCS was not successfully implemented anywhere. The UN's Intergovernmental Panel on Climate Change (IPCC), established in 1988 as the main international agency collecting and publishing data on climate change, continues to promote CCS but acknowledges it involves significant costs, reduced efficiency, and more energy inputs for power plants implementing the technology.[53]

Arguing about CCS

The IPCC has long emphasized that "no single technology option" can reduce emissions enough to reverse climate change, so "a portfolio of mitigation measures will be needed," of which CCS must be a part. In a major report in 2018, the IPCC proposed four potential pathways to limit warming to 1.5 degrees Celsius or less—the goal the IPCC argued was necessary in order to avoid "long-lasting or irreversible" impacts on humans and ecosystems. All require some form of carbon removal and significant investment in CCS technology.[54]

Many opinions in favor of CCS cite this 2018 report. The International Energy Agency argued that 15 percent of emissions reduction must come from CCS by mid-century. The groups that oppose CCS, supporters say, are ignoring the science.[55]

CCS is the path preferred by the fossil fuel industry, for obvious reasons. Not only does it promise to enable their continued expansion, but as an aspirational technology, it justifies massive federal investment in the industry. The labor movement is a strong advocate for investment in CCS and especially in CCUS technology. The industry-union lobbying group Carbon Capture Coalition supports continuing fossil fuel production as a source of good union jobs. According to the coalition's founders, "The industry's future growth depends on taking advantage of the large amounts of CO_2 that result from electricity generation and industrial processes," thus the coalition seeks "to encourage policies that will help bring an affordable supply of man-made CO_2 to the market."[56]

Some environmental organizations have signed on. While industry-heavy, the Carbon Capture Coalition also counts the National Audubon Society, the National Wildlife Federation, and the Nature Conservancy among its members as well as major agricultural, energy, and fossil fuel companies and unions.[57]

But CCS has plenty of detractors. Many environmental organizations argue that CCS development is a gift to fossil fuel companies, a pie-in-the-sky distraction from the real work of ending the use of fossil fuels.[58] Instead of pursuing ways to continue using fossil fuels, they say, we should be working to eliminate them, especially in sectors like power generation, where alternatives already exist.

Some of those who oppose CCS for use in electricity generation are willing to consider it for other uses, like capturing CO_2 emissions from industry or capturing CO_2 that's already been emitted from the atmosphere through a technique called direct air capture (basically giant fans that force air over substances that bind to CO_2). Direct air capture is known as a "negative emissions technology" in that it removes existing CO_2 from the atmosphere. Like CCS, direct air capture is still mostly aspirational and would require the development of new systems for storing captured CO_2.

Investing huge sums of money in new technologies that prolong an unsustainable and unjust system distracts us from the true problems: cutting fossil fuel use to reverse our growing emissions and transitioning to a just and sustainable system. The IPCC and other mitigation scenarios rely on "carbon removal technologies that barely exist," warned *MIT Technology Review* after the panel updated its recommendations in 2021. The IPCC reports studiously ignored the more obvious path: "Reducing energy and resource use," another important study concluded, "represents a safer and more ecologically coherent approach to climate mitigation." That, however, would require challenging the entrenched interests that benefit from the status quo.[59]

DOES LEED CERTIFICATION MEAN THAT OUR BUILDINGS ARE USING LESS ENERGY?

Standard statistics show that residential and commercial buildings account for some 13 percent of US GHG emissions, though this figure encompasses only on-site energy use. If we include the electricity that these buildings draw from the grid, residential and commercial emissions jump to 31 percent of total US emissions.[60] So reducing the emissions caused by this sector must be an important component of overall reductions.

The Leadership in Energy and Environmental Design (LEED) certification system claims to address this issue. But it's a classic example of good intentions being co-opted, so that the goal of reducing emissions is subordinated to a profit-making agenda. Many critics now see LEED as little more than greenwashing.

LEED grew out of an initiative from the American Institute of Architects (AIA) in the 1980s to reduce energy usage in the building sector by promoting more ecologically friendly building design. The AIA's new Committee on the Environment (COTE) received funding from the Environmental Protection Agency and sought to collaborate with building and development industry representatives. When COTE resisted being turned into an industry mouthpiece, real estate and building industry interests split off to create their own organization, the US Green Building Council (USGBC).

It was the USGBC that developed the Leadership in Energy and Environmental Design (LEED) standards in 1998. To obtain one of several levels of LEED certification, planners must fulfill a certain number of requirements on a LEED checklist. "A vast ecosystem of green commerce has grown in tandem with LEED," writes journalist Brian Barth. In fact, by 2018, green building had grown to a $1-trillion global industry.[61] In the process, it became less and less green.

There are basic structural weaknesses in the industry-designed LEED system. They begin with the fact that completing the checklist for certifying new building projects is a one-time process. Once a project obtains certification, it keeps it forever, with no follow-up to verify whether a building is actually meeting any energy or environmental standards in practice. Because there are no reporting requirements, it's very hard to evaluate the impact of certification on energy use.

A checklist system allows the specifics and minutiae of obtaining certification to overwhelm the larger environmental issues at stake. For example, my own university, Salem State, obtained Parksmart bronze certification from the USGBC, enabling it to declare a new four-and-a-half-story parking garage "green" and obtain tax-exempt state-issued "green bonds" to fund its construction. (Since the original LEED standards did not include parking garages, the council developed Parksmart, a set of comparable standards for such structures.) The project obtained the certification by including electric-vehicle charging stations and reserving spots for shared and energy-efficient vehicles.[62] But many, on and off campus, found it Orwellian to call a new parking garage green, when the green alternative would be to invest in infrastructure to help get people *out* of their cars.

But the biggest structural problem is that the standards were designed by industry itself. The USGBC is made up of the same businesses that it seeks to regulate. As an organization, its interest is in making money (a LEED certification costs $35,000 in fees) and helping building sector industries cash in on the tax and public relations advantages of a green seal of approval. In a study of the 7,100 successful certifications as of 2012, *USA Today* found that "designers emphasize LEED points that can be won through simple purchasing decisions and shun labor-intensive options and cutting-edge technology." In fact, the most popular checklist item was to include a USGBC-certified expert in the design team—a clear moneymaker for the USGBC, which runs the only certified training program. Several independent studies have shown that LEED-certified buildings are no more energy-efficient than their noncertified counterparts. Oberlin College physics professor John Scofield concluded, "What LEED designers deliver is what most LEED building owners want—namely, green publicity, not energy savings."[63]

LEED has also served as a way to siphon money from the public to the private sector. In addition to tax-exempt green bonds, many localities offer significant tax breaks to new buildings that meet LEED standards. So developers pay a one-time fee to the USGBC (a private, industry-run organization) and receive much more back in tax breaks—around $500 million for some two thousand LEED-certified projects. In fact, some government agencies and localities require new public construction to be LEED certified—at a cost to taxpayers of approximately $150,000 per project just for the fees paid to USGBC and LEED consultants. About 26 percent of LEED-certified buildings are government owned.[64]

CONCLUSION

Science and technology must play a crucial role in designing climate policy. But neither science nor technology alone can give us answers about how to restructure our society and our global and national economies to live within planetary boundaries. These are political questions to debate.

We need to use new technologies as we rapidly reduce our CO_2 emissions and our use of fossil fuels. But an ecological approach suggests that without accompanying political, social, and economic change, new technologies risk being absorbed or co-opted into our ever-growing, overconsuming economy, controlled by big business in the interest of profit. They also risk perpetuating, or exacerbating, the global divide between the haves and have-nots. The specific examples described in this chapter show how easy it is for polluting industries to adopt new technologies or standards as a form of greenwashing while continuing business as usual and making only a small dent in cutting emissions.

Despite increased public awareness and concern about the climate emergency and despite generations of scientific and technological advances, our planetary emissions continue to increase. We need to change our economy and our politics if we want to shift to a low-emissions society—our economy because we need to create a system of production that prioritizes human needs rather than increasing consumption, and our politics, because to make this change, we need to take control from a corporate system that has every interest in perpetuating itself.

CHAPTER 2

POLICY QUESTIONS

It makes sense to look at policy, as we did with technology, from a holistic or ecological perspective. We need to understand how individual policies interact with existing economic and political systems, and with other potential policy paths, and how national-level policies interact with global systems.

We've looked at how, in the absence of political change, technology tends to be absorbed into existing political and economic structures. At first glance, it seems governments addressing climate change on a policy level could bring the kinds of structural changes needed to reduce our use of fossil fuels and our emissions of greenhouse gases. The IPCC, and major international agreements like the Kyoto (1997) and Paris (2015) accords, agree that technology alone is not enough. Government action is necessary, and governments must implement policies to reduce emissions. But the world's most powerful governments are deeply invested in the economic status quo: an industrial fossil economy that fuels a profit system based on economic growth, which draws the poor as well as wealthy elites into its vortex.

This section looks at some of the ways existing government policies in the United States and the rest of the world exacerbate the climate emergency. It examines why the national and international policy actions, as of 2021, have failed to slow global emissions. An ecological perspective on policy can highlight some of the obstacles that have limited governments' capacity to imagine and implement the kinds of structural changes needed to alter our path towards accelerated environmental destruction.

This does not mean that policy should not play a role in addressing climate change. Indeed, the role of policy is key. But it means that governments are committed to existing systems that prevent them from enacting the kinds of radical policies we need to significantly reduce emissions. Without major pressure for change, governments are likely to continue to enact, at best, half-measures filled with exceptions that allow business—and emissions—to continue, with a few minor tweaks, as usual.

WHAT WAS THE KYOTO PROTOCOL?

By 1988, when the UN's Intergovernmental Panel on Climate Change (IPCC) was established, there was plenty of evidence that human activity, primarily through the burning of fossil fuels, was causing the earth to warm, and that it was happening faster than earlier predictions had expected. The IPCC brought together scientists from around the world to investigate what was occurring. Since then, the IPCC has released numerous reports documenting how human activity and GHG emissions are increasing atmospheric CO_2 levels and how the increase is affecting our climate and our planet.

At the 1992 Rio de Janeiro Earth Summit, most of the world's countries signed on to the United Nations Framework Convention on Climate Change calling on the world's industrialized, high-polluting countries to reduce emissions to limit dangerous human-caused climate change. The Conference of the Parties (COP) agreed to work towards an international treaty with binding targets to achieve this, culminating in the 1997 Kyoto Protocol, which established overall emission reduction goals.

One key area that the COP debated was the relative responsibility of different countries for historical and current emissions, and their differing responsibility for reductions. Should a poor country, with low energy use (and thus low emissions) be required to lower its emissions as much as fuel-guzzlers like the United States? Did wealthy countries have a responsibility to share resources and technology to support low-emissions economic development in poor countries?

Kyoto responded to this question by invoking the principle of Common But Differentiated Responsibilities (CBDR): all countries share a responsibility to act, but major differences exist among countries in their levels of emissions and in their economic and technological capacity to implement change. Industrialized countries have much greater current and cumulative contributions and much greater economic and technological capacities, thus it is their responsibility to lead in emission reduction.

Kyoto required all participants to submit annual inventories of their emissions and carbon sinks to the COP. Industrialized nations, including those in Europe, Japan, Canada, Australia, and the United States, also known as "Annex 1" or "Annex B" countries, committed to reducing their emissions by approximately 5 percent by 2012, relative to the base year of 1990. The rest of the world, those countries that contributed far less to global emissions, were not given targets but could participate through the Clean Development Mechanism (CDM). Over 150 countries ratified the protocol, as did the European Union as a region. It went into effect in 2005, with its first four-year commitment period to start in 2008. In 2012, the Doha Amendment established a second commitment period lasting until 2020. While scientists and climate activists celebrated, they cautioned that the targets were not ambitious enough to avert catastrophic climate change, that too many rapidly growing middle-income countries were exempted, and that there were too many loopholes that allowed rich countries to meet their targets without making significant reductions.

Flexibility mechanisms: Markets and offsets

Even as it required emissions reductions, the Kyoto Protocol created "flexibility mechanisms" that offered a way around painful cuts. A new international market in emissions permits turned them into "a new commodity."[1] Annex 1 countries could meet their targets by purchasing extra permits from other countries or obtaining credits for funding emission-reduction projects in poorer countries. The permit system created a financial incentive to reduce emissions, but also created a way to fulfill targets without actually lowering emissions.

Countries and regions implemented internal emissions markets as a flexible way of incentivizing different sectors to contribute to overall reduction targets. Instead of establishing firm quotas for different businesses, governments issued them emissions permits they could buy and sell. The rationale was that sectors, or countries, that could make low-cost shifts would do so quickly, and sell their permits to others for which changes would require greater investment or time. The ability to trade permits was supposed to bring the magic of market efficiency to the problem of reducing emissions.

Countries also purchased exemptions to their targets through offsets, like reforestation projects, or funding for supposedly renewable projects in the Global South through the Clean Development Mechanism (CDM). The CDM encouraged resource and technology transfers to help poor countries develop along a lower-carbon path. But its projects don't necessarily represent overall emissions cuts; they can even increase emissions at the global level.

Emissions markets and offsets have formed the basis for international climate policy ever since. I'll go into more detail on how these policies have worked later in this chapter.

The United States and Kyoto

For the United States, Kyoto's decision to require reductions only from rich countries proved fatal. Several months before the Kyoto meeting, the US Senate unanimously passed the Byrd-Hagel Resolution. It stated that "the United States should not be a signatory to any protocol . . . at negotiations in Kyoto in December 1997 or thereafter which would: (1) mandate new commitments to limit or reduce greenhouse gas emissions for the Annex 1 Parties, unless the protocol or other agreement also mandates new specific scheduled commitments to limit or reduce greenhouse gas emissions for Developing Country Parties within the same compliance period; or (2) result in serious harm to the U.S. economy."[2] Even climate progressives like Massachusetts's John Kerry and Minnesota's Paul Wellstone voted in favor of the resolution.

Still, the United States participated in the Kyoto meetings, imposing some of its priorities on the agreement, like pressing for targets softened with broad flexibility mechanisms and for the exemption of aviation, shipping, and military sectors. The US signed the final declaration under the Clinton administration. George W. Bush, however, declared his opposition during his 2000 campaign and announced definitively in 2001 that the United States would not ratify it. At the 1992 Rio Summit, then-president George H. W. Bush had announced that "the American way of life is not negotiable." His son seemed to agree.[3]

Canada pulled out in 2011, saying without the participation of the world's two largest polluters, the United States and China, the agreement was meaningless. (China ratified the protocol but was not designated an Annex 1 country, and thus was not subject to emissions reduction requirements.)

How effective was Kyoto?

The Kyoto Protocol required the developed countries to reduce their emissions by the end of 2012 to below 1990 levels. The Annex 1 countries did, collectively, reach their reduction targets. But Kyoto was not the only reason. The 2008 financial crisis and economic recession throughout the industrialized world caused significant drops in production. Also, the former Soviet republics suffered a deep recession in the 1990s after the collapse of the Soviet Union, further depressing parties' production and thus emissions. Ten of the thirty-six participants didn't reduce their emissions at all, achieving their targets with purchased carbon credits. Outsourcing (i.e., shifting industrial production to the

Third World) also contributed to reductions. Since aviation and shipping were excluded from the calculations, growth in these sectors due to increased trade was not accounted for.[4]

But with no consequences for countries that failed to meet their targets, without participation of the world's largest emitter—the United States, until China surpassed it in 2007—and with large countries like China and India exempted under the CBDR provisions, global emissions continued to rise inexorably.

WHAT IS THE PARIS AGREEMENT?

In the Paris Agreement, agreed on in 2015 and signed in 2016, the world's countries acknowledged the increased urgency of reducing emissions to limit global temperature increase to at most 2 degrees Celsius, preferably 1.5 degrees Celsius, above preindustrial levels. Signers agreed to create and submit plans to reduce their emissions to nationally determined targets compatible with that goal. Paris went beyond Kyoto by requiring all countries, not just the most industrialized, to submit plans for reductions. Like Kyoto, Paris relied on flexible mechanisms and carbon markets and created a new carbon offset program, Reducing Emissions from Deforestation and Forest Degradation (REDD). And Paris was weaker than Kyoto in that it did not mandate specific targets; every country was asked to set its own.

Significantly, Paris required every country report on its carbon emissions in great detail. This kind of information is crucial for trying to understand the contours of the problem and implement policy steps to reduce emissions.

The Paris Agreement, five years in

The Paris Agreement's greatest strength—bringing all countries on board by letting them set their own targets—is also its greatest weakness. Many countries submitted plans nowhere near strong enough to meet even the 2-degrees Celsius goal, with no consequences for failing to fulfill their own plans. Only two countries, Morocco and the Gambia, submitted plans consistent with meeting the 1.5-degrees Celsius target. None of the world's largest emitters, which were mainly the wealthiest countries and China, came close to reductions compatible with the agreement's goals. The United States withdrew from the agreement between 2019 and 2021, under President Trump. And global emissions continued to rise, despite a temporary drop in early 2020 with the COVID-19 pandemic. By December of that year, emissions had more than recovered, leaving the balance for 2020 at 2 percent over 2019's emissions.[5]

"If all governments meet their Paris Agreement target, we calculate the world would still see 3°C of warming, but that warming is likely to be even higher given most are not taking enough action to meet their targets. We still have a long way to go," warned Bill Hare of Climate Analytics in late 2019.[6]

The IPCC announced, in a 2018 special report, that countries' announced targets, along with ongoing emissions trends, had the world on track to significantly exceed the warming goal of not more than 2 degrees Celsius. "Without increased and urgent mitigation ambition in the coming years, leading to a sharp decline in greenhouse gas emissions by 2030, global warming will surpass 1.5°C in the following decades, leading to irreversible loss of the most fragile ecosystems, and crisis after crisis for the most vulnerable people and societies." The IPCC concluded that while "limiting warming to 1.5°C is possible within the laws of chemistry and physics," it "would require unprecedented transitions in all aspects of society" that none of the world's large emitters seemed prepared to confront.[7]

WHAT ARE THE ADVANTAGES AND DISADVANTAGES OF PUTTING A PRICE ON CARBON?

Absent the political will to actually ban or heavily regulate fossil fuels and emissions, many governments have relied on carbon pricing systems to encourage emission reductions by creating financial incentives. They turn permission to pollute into another cost of doing business. Corporate and financial sectors tend to favor trading systems because they create new business opportunities. Trading systems and carbon taxes leave business decisions in the hands of the polluting industries, so they do not entail significant disruption to the economic status quo.

Some environmentalists support carbon pricing, arguing industries that create emissions have historically been able to evade responsibility for the costs their emissions incur on society as a whole. Putting a price on emissions forces companies to pay part of the cost themselves. Many also believe that it's the most practical way to lower emissions quickly.

Even among supporters of carbon pricing, there is debate about the specifics. Some think carbon pricing should be the main approach to lowering emissions, others see it as a component of a larger overall plan. Many analysts and grassroots organizations believe existing carbon pricing programs serve as a distraction that allows fossil fuel companies and associated industries to continue business as usual. Companies can continue to pollute, while compensating for their new expenses by reallocating resources, raising prices, or lowering costs by abandoning workers and communities.[8]

From a climate justice perspective, an effective carbon pricing mechanism must fulfill two goals. One, it must be accompanied by an enforceable, science-based and rapidly declining cap, to ensure that emissions actually decline fast enough to avoid catastrophic climate change. And two, it must be redistributive, rather than increasing inequality.

HOW DOES A CAP-AND-TRADE SYSTEM WORK?

Some countries and regions have introduced internal cap and trade or Emissions Trading System (ETS), modeled on Kyoto's system. A national, state, or regional government establishes an annual cap on CO_2 emissions from particular types of carbon-producing entities. It then grants or auctions carbon-emitting permits, or "allowances," to individual players to keep total emissions within the national cap. In most systems, the cap is designed to go down year by year, encouraging industries to plan for ongoing emissions reduction measures. The "trade" element introduces flexibility into the process: because they have the option of purchasing extra permits, companies make their own decisions about how to factor the cost of polluting into their business models.

Most systems allocate a certain number of free allowances and auction the rest. There's a limited number of permits (hence, the "cap"), so as the supply dwindles, the price will rise, creating more incentive to reduce emissions. Companies basically buy and sell, or "trade," the right to pollute, under a nation- or region-wide limit.

The rationale for the trading system is that some industries have easier paths to reducing emissions than others. Trading encourages reductions to take place quickly, where it's less onerous, and gives other businesses more time to make the transition to low-carbon technologies. As caps go down and allowances become more expensive, the cost of emitting CO_2 will rise, and market incentives will push more businesses to reduce more.

The European Union has operated the world's first and largest region-wide ETS since 2005. In the United States, the Northeast Regional Greenhouse Gas Initiative (RGGI) went into effect in 2009, followed by one in California in 2013. China launched the world's largest nationwide ETS in January 2021.

While all Emissions Trading Systems follow the same basic principles, they vary as to how stringent the caps are, whether free allowances are distributed and if so, how many, how prices for allowances are set, what sectors they apply to, and how the income from allowance sales is used. Allowance trading creates new private financial markets, while government auctions create revenues that may be used for particular purposes like investing in energy transitions or returned to the public through a dividend system.

All of these variables bear on the potential effectiveness of the system. If caps are high, allowances are cheap, and some sectors are exempt, the system may have little impact on emissions. If caps are low and allowances costly, the costs can incentivize businesses to seek lower-emissions alternatives in their production processes. The most effective option would be a hard and declining cap that covered all emitting sectors (or was imposed on fossil fuel extraction), with no loopholes. It could be science-based, progressively lowered to set us on the 1.5-degrees Celsius path, and global. But most existing systems have "soft caps" with so many escape hatches that even if some emissions go down, it hasn't affected the bigger picture.

The EU system caps emissions from power plants, some industries, and intra-European airline flights. It applies to some eleven thousand power stations and factories that, along with the flights, produce 45 percent of the EU's GHG emissions. The cap goes down every year to ensure that emissions decline. Companies receive free individual allowances that likewise decrease annually, and they can participate in a region-wide auction to purchase further emissions credits. They can also "bank" permits for future use. (Airlines still receive most of their allocation free.) Companies can also purchase credits by investing in CDM projects. Those that exceed their allowances are heavily fined.[9]

Emissions in the EU's covered sectors indeed went down, projected to be 21 percent lower than 2005 levels by 2020, and 43 percent lower by 2030.[10] But some analysts say the ETS had little to do with the decline. The collapse of industry in former socialist bloc countries and the 2008 recession played a major role. The ETS coincided with a decline in Europe's use of coal, the most polluting of all the fossil fuels, in favor of natural gas. But falling natural gas prices may have played a bigger role in the shift.

Critics point out that 80 percent of Europe's manufacturing industry and 85 percent of the aviation industry are still receiving free credits, while most industrial, residential, and commercial emissions are not even covered by the ETS. Even in covered sectors, carbon was being priced far too low everywhere to significantly affect business decisions.[11] The 2008 recession allowed companies to bank credits, pushing prices down and further weakening the system's financial incentives for reducing emissions.

The RGGI in the US Northeast is in some ways stronger than the EU's ETS, even though it's narrower in scope, applying only to electricity-generating plants. Unlike Europe's system, the RGGI does not distribute free allowances. Revenues from allowance sales—some $3.9 billion by 2019—are invested in energy efficiency and other emission-reduction projects. As also occurred in Europe, the 2008 recession and low natural gas prices encouraged a shift away from coal that contributed to lowered power plant emissions during the system's first decade. But it's notable that power plant emissions declined significantly *more* in the RGGI states than in the rest of the United States, which suggests that the RGGI played a role.[12]

Nevertheless, energy journalist David Roberts concluded that the cost of allowances "acted as a small carbon tax, but not one large enough to shape broader market dynamics." Both supporters and critics note that the RGGI's limited scope—applying only to electricity generators—restricts its ability to significantly reduce CO_2 overall emissions.[13]

California's cap-and-trade system, which went into effect in 2013, applies to large industrial plants and fossil fuel distributors as well as power generators. Like the EU system, California's system grants free allowances, and like both systems described above, it allows banking of allowances and has consistently maintained a surplus and kept allowance prices low. Oversupply and banked allowances make the cap somewhat moot, since covered installations can continue to use their banked allowances to exceed future caps. The state's powerful oil industry managed to negotiate another of its goals into the legislation: a prohibition on any further regulation of CO_2 emissions.[14]

Emissions from California's oil and gas industry actually *rose* during the first six years of the state's cap-and-trade system. Only in the electricity sector—where the state cut its imports from out-of-state coal plants—did the added cost seem to motivate change. Overall, California industries subject to regulation lowered emissions more than those subject to cap and trade.[15]

Putting a price on carbon might be a useful component in a broader program to lower emissions. But thus far, governments have been too accommodating to industry, setting caps too high and prices too low and even agreeing, as in California, to withhold other forms of regulation. The World Bank, which supports carbon pricing as a step towards lowering emissions, warned that less than 5 percent of the fifty-seven national and subnational carbon pricing systems in place as of 2019 had set their rates high enough to significantly affect emissions.[16]

To stronger critics on the left, carbon trading leaves too much power in the hands of major polluters that got us into the problem in the first place, allowing them to make even more money through new financial schemes while evading government action to legislate change, regulate, or even tax them. In fact, many systems *reward* the big polluters by granting permits to pollute based on their historical pollution.

For many critics, the emphasis has been far too much on the "trade" aspect of so-called cap-and-trade systems. Pope Francis articulated this position when he argued that "the strategy of buying and selling 'carbon credits' can lead to a new form of speculation which would not help reduce the emission of polluting gases worldwide. This system seems to provide a quick and easy solution . . . but in no way does it allow for the radical change which present circumstances require. Rather, it may simply become a ploy which permits maintaining the excessive consumption of some countries and sectors."[17]

Even most cap-and-trade critics support the concept of capping emissions but contextualized quite differently from existing trading systems. Instead of letting the market nudge polluters to reduce in some sectors, we need universal hard caps to force emissions to go down in accordance with the trajectory we need to limit warming to 1.5 degrees Celsius: about 7 percent a year.

The most direct way is to impose the cap at the source—on fossil fuel extraction.

"Cap and share" or "cap and dividend" proposals add the principle that polluters must compensate the public for the harms they cause. In the strongest versions, the cap would be global. Instead of distributing free permits within the cap, governments would auction them, returning the proceeds to the population through a dividend. As the cap declines and the cost of permits rises, fuel producers might raise their prices, but the dividend would also increase, easing the burden on consumers. The uniform return would be proportionally more significant to poorer people.[18] No such experiment has been implemented, much less on a global level, but it does suggest an alternative for making caps more effective and fairer.

HOW DOES A CARBON TAX WORK?

A carbon tax constitutes a more direct approach to making emissions more expensive and obligating emitters to pay for part of the harm they cause. The government calculates the carbon emissions of different activities and levies the tax according to the emissions created. Like ETS, the tax enables businesses to incorporate new costs into their planning and make their own decisions about the cost-effectiveness of reductions. A tax can be designed to lower emissions over time by going up regularly, encouraging producers to invest in long-term CO_2-reduction measures.

Proponents say a tax is simpler and more direct than cap-and-trade schemes. Rather than creating complex new bureaucracies and financial markets, it relies on existing government systems. But unlike cap and trade, the tax alone creates no "cap," only an economic incentive.

Using taxes to influence markets is not new. Taxes on cigarettes and alcohol are based on the premise that by making these products more expensive, the government can discourage their use, while still leaving the ultimate decision in the hands of the consumer. A carbon tax makes even more sense from this perspective, because the activity is harmful not just to the person who uses the product but to society in general. Tax revenues can also be used to compensate society for the harm that users cause.

Such a tax can be imposed at different points in the extraction, production, importation, sale, and use of fossil fuels. If the tax is imposed on fuel extraction, production, or importation, it's categorized as supply-side or upstream; if it's imposed on sale and consumption, it's a demand-side or downstream tax. Demand-side taxes, like that for gasoline imposed at the pump, tend to provoke more public opposition, as they are more immediately visible to the consumer. Either way, though, such taxes can be factored into the costs of business and passed along to consumers.

Unlike with alcohol and cigarettes, consumers often don't have a lot of choice about basic consumption of fossil fuels and their products. People need to eat and, depending on the weather, heat or cool their living spaces. Many depend on their cars for essential transportation. When food, utility, or gasoline prices go up, it's the poor that suffer most, and they often protest vociferously.

Like any policy, a tax can incorporate loopholes and exemptions, which businesses generally lobby hard for. Argentina's 2018 carbon tax, for example, excluded fuels for international aviation and shipping, as well as fuel exports. Canada's exempted the steel and chemical industries.

Existing taxes, analysts agree, are far too low. Like existing cap-and-trade systems, they have contributed to hastening the transition from coal to natural gas but are unlikely to have much impact beyond that. "In practice, most countries have found it politically difficult to set prices that are high enough to spur truly deep reductions," concluded the *New York Times* in 2019, looking at the over forty countries or regions that had imposed some form of carbon pricing.[19]

The Rhodium Group found that even with a relatively higher tax than has yet been attempted, beginning at fifty dollars per ton in 2020 and increasing 2 percent a year, its impact would be mostly in the electricity generation sector, where cheaper alternatives are available, and would occur during the first ten years. Beyond that, more radical measures would need to complement even a high tax.[20]

How industry came to love the tax

In a landmark 2007 decision, the US Supreme Court allowed the Environmental Protection Agency to consider greenhouse gases a pollutant under Clean Air Act provisions, if it found they endangered public health and welfare. The Clean Air Act offers one of the strongest regulatory mechanisms around; it has been called "the world's most successful pollution reduction program."[21]

In 2009, the Obama administration declared that GHG emissions did indeed threaten public health and welfare, and the EPA began to implement a series of emissions-control regulations, mostly aimed at power plants. In 2015, these were solidified in the Clean Power Plan, which would have capped power plant emissions with enforceable limits in 2022 and reduced them to 2005 levels by 2030. Combined with federal promotion of natural gas extraction, Obama's policies accelerated the shift from coal to natural gas in the US electricity generation sector.

The Trump administration worked quickly to dismantle these measures. But given the new infrastructure already put in place and low natural gas prices, the coal industry did not benefit much. Coal plants continued to close.[22]

As states and regions began to implement their own policies in the absence of federal action, the fossil fuel industry confronted the challenge of having to comply with multiple complex systems. Just as the automobile industry advocates national fuel efficiency standards to avoid a plethora of different state regulations, so the fossil fuel industry prefers a single standard. No industry wants to be taxed. But a uniform tax was better than a maze of different rules and regulations.

Moreover, the oil and gas industries are well placed to gain more than they'll lose from measures that increase the cost of emissions, unlike coal companies that will suffer at the expense of natural gas. Oil producers expect to feel little effect from a tax; with no ready alternative to their product, they can simply pass along any added costs to consumers.

Moreover, as public opinion and activism have recognized the climate emergency and demand action, the industry wanted to burnish its image and, as with cap-and-trade systems, secure a seat at the table as policies are designed. Its priorities: the tax must not be too high and must be paired with restrictions on regulatory measures and provisions that protect the industry from lawsuits. On a populist note, the industry joined with fiscal conservatives and progressives to promote making a tax revenue-neutral, returned to taxpayers in the form of dividends. Conservatives advocated the dividend because they generally oppose raising taxes and government spending on social services, while progressives supported the dividend system because of its redistributive effects.[23]

Climate activists often advocate a higher tax, or one that will go up over time, and oppose the restrictions on regulations and lawsuits. They're more divided on the dividend. Some see it as a way to redistribute wealth more fairly, while others think it should be invested in the transition to a low-carbon future.[24]

In 2017, the Climate Leadership Council (CLC) brought together a broad coalition of environmental organizations as well as mining and energy companies, automobile manufacturers, high tech, insurance, other major manufacturers, and financial industries to press for a carbon tax and dividend system that "embodies the conservative principles of free markets and limited government." Its plan for a gradually rising tax on emissions, and a companion fee on imports, garnered bipartisan support in Congress. The CLC proclaimed itself as "Pro-Environment,

Pro-Growth, Pro-Jobs, Pro-Competitiveness, Pro-Business and Pro-National Security."[25]

The CLC doesn't advertise what *Bloomberg* called the plan's most "potent prizes"—a prohibition against climate-related lawsuits tied to past emissions and an end to federal regulations on emissions. It also offers "a favored business alternative to the more aggressive Green New Deal."[26]

Taxes and social justice

No matter how tax proposals have been framed, they've been hard to sell to the public. Carbon taxes can and generally will be passed along to consumers. (If it's a consumption tax, like a tax on gasoline, it's paid directly by the consumer.) And consumers are resistant to policies that make energy costs higher. Poor people spend a greater portion of their income on energy costs, whether for heating their homes or filling their vehicle's tank with gasoline. Rising fuel costs have frequently been a spark that has driven populations to the streets to say "no."

In France, government attempts to raise fuel and energy costs in 2018 were met with widespread public protest by the *gilets jaunes*, or yellow vest movement. In Ecuador too, public outrage forced the government to cancel a proposed fuel tax. In Washington State, voters rejected two carbon taxes, in 2016 and 2018. It's worth looking into these policy attempts, and the public reaction, to consider the social impacts and political viability of this type of climate policy.

In 2018, French president Emmanuel Macron announced new fuel taxes adding about twelve cents to the price of a gallon of gasoline, and twenty-four cents to a gallon of diesel. Protests erupted around the country, first in rural areas and then in Paris and other cities. Protesters identified themselves with yellow vests, which all drivers in France are required to carry in their vehicles for emergencies.

The protests spread rapidly through social media, apparently not organized by any major political party or movement. Within weeks, hundreds of thousands were taking to the streets, blockading highways, burning vehicles, and smashing luxury shops in Paris. The government soon backed down and rescinded the tax.

While Macron presented the tax as part of an environmental agenda—he had vowed to eliminate fossil-fueled vehicles altogether by 2040—the public tended to view it in the context of his campaign to shift tax burdens from the wealthy to the poor and middle class. The tax exempted most businesses and industries, exacerbating a public perception that it was aimed at the poor. It also disproportionately affected rural France, where less access to public transit and municipal services mean people rely more on their cars and directly purchased heating fuel. Polls showed that over 70 percent of the population supported the movement, if not all of its tactics, while Macron's popularity dropped to 23 percent during the crisis.[27]

In Ecuador, President Lenín Moreno also faced widespread and sometimes violent protests in late 2019 after his government announced an austerity package sponsored by the International Monetary Fund (IMF) that included an end to fuel subsidies. The country's powerful Indigenous organizations led the protests that followed as fuel prices doubled. The government declared a state of emergency before finally rescinding the cuts.

Moreno did not present his cuts in environmental terms. And popular protest against IMF-sponsored austerity programs that weigh most heavily on the poor have a long history in Latin America.

As these examples show, strictly market-based measures can worsen inequalities. The costs will be experienced immediately by the poor, while the potential benefits seem abstract. And the population may refuse to go along.

The two Washington State referendums, in 2016 and 2018, offered models for making a carbon tax more attractive—or at least more palatable—to voters. The first, like the CLC proposal, would have returned revenues as a dividend. The second would have invested revenues in carbon-reduction alternatives and public services that would both decarbonize and redistribute. Neither referendum passed.

Some climate radicals reject the idea of a tax altogether as implicitly legitimizing the status quo, leaving decisions in the hands of industry and enabling business as usual. A tax implies that while its object may be a luxury, or socially negative, it should be allowed to continue. To make a tax truly climate-effective, some argue, it must be combined it with the kind of hard and declining cap described above.

WHAT ABOUT CARBON OFFSETS?

The idea of carbon offsets has been around since the Kyoto Protocol's Clean Development Mechanism (CDM). Offsets give companies, or countries, a way to compensate for their emissions by paying to preserve carbon-consuming forests or by subsidizing supposedly low-carbon alternatives elsewhere, especially in the Third World.

The CDM acknowledged the fact that poor countries needed to develop energy production, and that rich countries monopolized the resources and technology necessary to allow this to happen in a low-carbon manner. The CDM thus encouraged rich countries to make such investments to offset their domestic emissions. Other offset programs involve preserving or planting forests to, supposedly, remove CO_2 from the atmosphere. Some projects enable individuals to purchase offsets to make up for high-emissions expenditures, as when airlines encourage passengers to purchase offsets along with their plane tickets.

Carbon offset systems alone don't impose any cap, but they often complement cap-and-trade or national target systems by helping entities meet their goals without necessarily lowering their emissions. Industries tend to support offset options because they give companies control over the process, with little oversight. And they are great for public relations.

But reforestation and forest preservation schemes are plagued by weak definitions of "forests," lack of regulation, and leakage.

Most rely on the UN Food and Agriculture Organization (FAO) definition of forests, which includes "forest plantations"—big agribusiness operations destined for logging, biomass, rubber, and other uses, even when it means clearing existing forest in order to plant new trees. Lack of regulation also means that "forests" planted or preserved one year might be deforested or harvested (i.e., cut down) the next. Offset projects are also undermined by what's known as leakage—designating one site for protection simply transfers logging or other deforestation activities to a different site.[28]

The definition of "clean" energy is similarly vague; it has even been stretched to include new coal-fired power plants that are more efficient than older ones.[29]

The international organization Carbon Market Watch, while not completely opposed to the idea of offsets, concluded that the experiments thus far have been a failure. "It is often impossible to know where the money goes to and if it is actually used to reduce carbon pollution. . . . The Kyoto's Clean Development Mechanism (CDM) has not reduced emissions and projects funded under it have been linked to human rights violations and environmental destruction."[30]

The CDM was supplemented in the Paris Agreement with Reducing Emissions from Deforestation and Forest Degradation (REDD). The Coalition for Rainforest Nations, led by Papua New Guinea and Costa Rica, initially proposed REDD to the UN Framework Convention on Climate Change in 2005. Third World countries would propose forest preservation projects, while industrialized countries would provide the funding.

Indigenous peoples and forest dwellers have decried REDD's lack of acknowledgment of their rights. Tom Goldtooth of the Indigenous Environmental Network protested that REDD "uses forests, agriculture and water ecosystems in the Global South as sponges for industrialized countries' pollution, instead of cutting emissions at source. . . . [REDD] may result in the largest land grab in history. It steals your future, lets polluters off the hook and is a new form of colonialism." The No REDD in Africa Network concurred: "REDD and all its variants are unjust mechanisms designed to usher in a new phase of colonization of the African continent. The REDD mechanism is a scam and the polluters know that they are buying the 'right' to pollute. . . . REDD provides a perfect space for polluters to keep polluting while claiming they are champions of climate action." In Panama, Indigenous groups fought successfully for the country to withdraw from participation in the program.[31]

Carbon offsets from the bottom up

A different kind of proposal, outside of the CDM and REDD systems, came from Ecuador, which incorporated the concept of the "rights of nature" in its 2008 Constitution. As other Latin American countries were relentlessly pursuing mining and extractive enterprises, Ecuador held a national referendum in which people voted to permanently forbid extraction in the Yasuní National Park, a rainforest home to at least two uncontacted Indigenous groups and others living in relative autonomy. The Yasuní project, unlike other carbon offset programs, would directly prevent the development of a significant fossil fuel deposit.

The oil fields in question contain close to a billion barrels of oil, some 20 percent of the country's reserves. In return for part of the income the country would forego, the United Nations agreed to establish a trust to receive international funding of $350 million a year over thirteen years. Pledges, mostly from a few European and Latin American countries, never came close to the goal.

In August 2013, Ecuador's president, Rafael Correa, announced the failure of the international commitment, with only $13.3 million delivered to his country. He terminated the program, liquidated the fund, and opened the land to drilling.[32]

Although Ecuador's project was not a traditional offset project—it didn't grant donors the right to pollute in exchange for their support—its collapse nonetheless illustrates the problems raised by Indigenous and Third World peoples. Their rights, and the rights of nature, matter little to global financial interests.

HOW DOES THE UNITED STATES SUBSIDIZE THE FOSSIL FUEL INDUSTRY?

The United States and other countries subsidize the fossil fuel industry directly, and in many indirect ways. Estimates vary widely, because of differences in how "subsidies" are defined. The United States gifts fossil fuels about $26 billion a year in direct budgetary transfers, tax breaks, grants and loans for research and development, and funding and insurance for overseas projects. Both Democrat and Republican administrations have promoted fossil fuels in the name of economic growth, jobs, competitiveness, and energy independence.[33]

Tax benefits that the industry enjoys include deductions for the cost of drilling new wells, deductions for depletion of reserves, tax credits for coal gasification and carbon sequestration projects, and tax credits, until 2014, for domestic energy production projects. The Department of Energy provides specific grants and loans for fossil fuel research and development. Finally, the Overseas Private Investment Council (OPIC) and the Export-Import Bank (EXIM) finance and insure US energy companies operating abroad. During Obama's supposedly climate-friendly administration, EXIM alone spent nearly $34 billion funding some seventy fossil fuel projects around the world.[34]

Many of these subsidies are particularly dangerous because they facilitate investment in new fossil fuel development projects, just when we should be working to halt these projects and shrink existing projects. Instead, as the authors of a recent study write, "Subsidies can beget more subsidies, with new, long-lived fuel infrastructure in turn (1) requiring further subsidization down the line to continue operating, and (2) yielding beneficiaries who will vigorously defend continued subsidization. Since there can be a revolving door between government staff and subsidy recipients, public officials may find it even harder to pass strong climate and energy policies. Indeed, the most troubling impact and legacy of fossil fuel subsidies may be the political barriers that fossil fuel producers have erected in recent decades against decarbonization efforts."[35]

Direct subsidies are just the tip of the iceberg. Through indirect subsidies, the government helps companies keep their costs artificially low by assuming public responsibility for the damages that fossil fuel use imposes on society. Governments (i.e., the public) assume the costs of pollution, GHG emissions, and accompanying social harms of fossil fuel use—an estimated $5.2 trillion a year globally. Including these indirect subsidies, the IMF found that the United States was gifting the fossil fuel industry some $650 billion a year, second only to China ($1.4 trillion), followed closely by Russia and the European Union. Almost half of these global subsidies went to the dirtiest fuel source of all—coal.[36]

Even the IMF definition leaves out other kinds of indirect subsidies. When the government promotes fossil-fuel-dependent industry and agriculture, funds fossil-dependent infrastructure like highways and suburban housing developments, and encourages fossil-dependent consumption and fossil-dependent economic growth, that's another kind of indirect subsidy. When it passes corporate-friendly trade deals, contracts military supplies and aid, and goes to war, those actions also bring windfalls to the fossil fuel industry. These kinds of subsidies are so deeply integrated into every sector of our economy that they are extremely hard to measure.

Although numerous agencies and governments have called for a phase-out of fossil fuel subsidies, in fact the opposite is happening. Counterintuitive as it might seem, we're actually *increasing* public funding for the industry. Not until early 2021 did the United States and the European Union begin to hint at cutting some of these subsidies.

The 2008–09 bailout of the US automobile industry reveals how deeply the notion of our economic well-being as a country, and thus our public policy, is tied to fossil fuels. Presidents Bush and Obama, and many economists, agreed it was in the national interest that the automobile industry survive the 2008 recession. Hundreds of thousands of jobs depend directly on the industry, as do other industries, like parts suppliers. One executive argued, "If GM and Chrysler would've gone into free-fall, they could've . . . taken the U.S. from a recession into a depression." Like the country's large banks, the auto manufacturers seemed "too big to fail."[37]

Rather than continue to subsidize and promote an economic system that is inherently unviable—our current fossil-dependent system—government resources and policy *should* be geared towards implementing and managing a rapid transition away from fossil fuels. It's true that under our current system, the elimination of fossil fuel subsidies could have dire impacts on people who are already marginalized. That's why a lot of people—and some of the proposals outlined below and later in this book—emphasize the need to think radically and holistically about changes to the system itself.

IS NATURAL GAS A BRIDGE FUEL?

To understand the impact of natural gas in our energy mix, we need to acknowledge apparently contradictory facts. Natural gas is a fossil fuel and a major source of GHG emissions. The more natural gas we burn, the more emissions we create. On the other hand, emissions from burning natural gas are only about half of those from coal. So if a country, or region, state, or individual power plant, switches from coal to natural gas, its emissions will decrease significantly. In fact, this shift has been a major factor in emissions declines in the United States and other wealthy countries in the past decades. President Obama called natural gas the "'bridge fuel' that can power our economy with less of the carbon pollution that causes climate change" as we transition to even cleaner sources.[38] But instead, natural gas extraction and use, and thus its emissions, have just increased.

Natural gas has traditionally been extracted by oil companies like Exxon, Chevron, and BP, from the same deposits as petroleum. Its production took a leap around 2007 with the expansion of hydraulic fracturing, or fracking, using high-pressure injections of water, sand, and chemicals to fracture shale rock and release the contained gas. Because of the amounts of water and toxic chemicals the process requires, and its contribution to deforestation, desertification, groundwater contamination, methane emissions, and other local and climate pollution, many environmentalists oppose fracking, even if they support increased use of natural gas.

In an American Petroleum Institute promotional video, President Obama says that "the natural gas boom has led to cleaner power, and greater energy independence" and pledges to "speed up oil and gas permits." "Natural gas isn't just appearing magically," Obama explained. "We're encouraging it and working with the industry."[39]

Public investment in fracking technologies actually started well before Obama. The Department of Energy explained proudly, "The increase in shale oil and gas production in the United States follows many years of investment and research carried out by the federal government. . . . From the 1970s to the 1990s, several DOE-funded R&D technologies would optimize production of shale across the United States. . . . These investments—combined with industry collaboration—made the American shale revolution possible."[40]

While Obama imposed some emissions regulation on the natural gas industry, most of it was rescinded by the Trump administration. During his presidential campaign, Joe Biden made contradictory remarks, variously suggesting that he did not oppose fracking, would restrict new fracking permits, and opposed fracking on public lands.[41] Once in office, Biden quickly announced some high-profile measures against the gas industry, including a ban on new public lands fracking and cancellation of the controversial Keystone XL pipeline permit. He also vowed to restore Obama-era regulation on methane leakage.

At the same time, Biden promised unions that he was "all for natural gas." His touted ban did not affect the 90 percent of fracking that takes place on private or state-owned land, and it did not affect existing projects. President Trump approved 4,700 new leases in his last year in office—1,400 in the last three months, ensuring that companies would be drilling on federal lands for many more decades. And Biden did nothing to slow down the permitting of new projects on these existing leases.[42]

Methane, the hidden GHG

Promoters of natural gas emphasize the fuel's lower CO_2 emissions, compared to coal, when it's burned to produce electricity. They overlook the leakage of methane—eighty times as potent as CO_2—in the fuel's production and transport. Methane leakage is not tallied in any systematic way, so estimates vary widely, and its omission from government statistics distorts its impact on the climate.

The methane that leaked from a 2018 accident at an ExxonMobil natural gas well in Belmont County, Ohio, never appeared in reported statistics. But a *New York Times* study using satellite remote sensing technology showed it was one of the largest leaks in history; it released more methane than the oil and gas industries of Norway or France emit in a year. "Methane emissions from oil installations are far more widespread than previously thought," the study concluded. Atmospheric measurements of methane show that its presence spiked after 2007 with the fracking boom. One study concluded that shale gas accounted for one-third of the increase in global methane emissions since 2007.[43]

Some scientists estimate that 2.3 percent of the natural gas we produce and transport goes directly into the atmosphere through methane leakage. Others believe it's as high or higher than 3.5 percent. At a 3-percent leakage rate, natural gas provides no advantage at all over coal, in terms of climate impact. Given a 2.3-percent leakage rate, "natural gas power plants are slightly more beneficial to the climate in comparison to coal-fired plants."[44]

A long bridge
If we could convert existing infrastructure from burning coal to burning natural gas while we develop a cleaner energy system, the idea of a bridge might make sense. But that's not the case. Switching from coal to natural gas means investing in new infrastructure, from fracking operations to processing plants, pipelines, and gas-burning power plants. This investment sets a "path-dependency" or "carbon lock-in." That is, once we've built the infrastructure, it's cheaper to keep using it than to go back and choose a different path. Natural gas is more like a bridge to more natural gas than to some other form of energy.

Since 2015, CO_2 emissions from natural gas in the United States have exceeded those from coal. US exports of natural gas, both via pipeline and in the form of liquified natural gas (LNG), have also surged, and massive new liquefaction and export facilities are being constructed. After record highs in 2018, exports doubled again in 2019.[45]

Policies promoting (and subsidizing) the extraction of natural gas are increasing its use worldwide, "fueling the global growth in greenhouse gas emissions." Natural gas can only lower emissions if it replaces coal, not if it increases fossil fuel use. The idea of natural gas as a "bridge fuel" only works if the bridge is very short. A 2021 study found the United States would have to drastically reduce its use of natural gas by 2030 to meet President Biden's announced emissions target by that year. Yet the American Petroleum Institute projected that oil and gas would continue to supply close to half of global energy needs through 2040, even if their use in the United States declined.[46]

HOW HAVE THE FOSSIL FUEL AND OTHER INDUSTRIES INFLUENCED POLICY DISCUSSIONS?

The fossil fuel industry has intervened on many levels to shape public discussion on climate change in the United States and worldwide. It's been the major force behind climate denialism: the argument that climate change (a) is not occurring at all, (b) is not caused by human activity, or (c) is not harmful. It's sought to shape legislation in its favor, in particular by seeking protection from regulations and lawsuits and by advocating subsidies and market-based policies that can create new avenues for profit. Perhaps most subtly and most effectively, it's helped shape how ordinary people think about the intersection of climate change with the economy—production, consumption, and jobs. It's encouraged us to conceive of climate change as a problem solved by technical fixes and individual consumer decisions. Industry will help by offering ever-more products and options. And it encourages us to see fossil fuels as the source of jobs, prosperity, and security.

Denialism

In *Merchants of Doubt*, Naomi Oreskes and Erik Conway explained that just as the tobacco industry funded studies to cast doubt on the relationship of smoking and lung cancer, so the fossil fuel industry fostered, funded, and promoted a few maverick scientists who denied the reality of climate change. With the advertising and public relations savvy of industry behind them, their voices were able to enter the public sphere and the mainstream media. "In creating the appearance of science, the merchants of doubt sold a plausible story about scientific debate," Oreskes and Conway wrote. The media, obligingly, responded with a commitment to "balance" and to reporting "both sides" in an evenhanded way. Even as internal communications at companies like Exxon made it clear that they understood the impact that fossil fuels had on the climate, their public-facing communications were steeped in denial.[47]

Climate denial campaigns succeeded in sowing doubt among the population. As recently as 2010, almost half of the US public believed that "the seriousness of global warming" was "generally exaggerated." By 2019, though, about two-thirds of the US population understood that climate change was indeed occurring, that it's caused by humans, and that its effects are harmful.[48] Still, that meant that a third of the population continued to doubt. And even as denialism lost ground, the country in 2016 elected a president who espoused it.

Although denialism persists, the industry has mostly given up that battle. Now it's moved on to greenwashing, techno-optimism, and an economic development narrative. Fossil fuel companies tout their investments in "clean" new technologies and research. All of them emphasize their commitment to the environment and the role of fossil fuels in advancing the interests of poor and marginalized communities worldwide. Rather than directly deny, they avoid mentioning the disasters that climate change is wreaking on the world's poor.

A 2018 study traced the evolution of oil industry positioning:

> At first with the support of the think tank network, Big Oil denied climate change. Then, in a second delaying tactic, it switched to advocating voluntary actions by individual companies and carbon markets. Finally, in 2007, it decided on the way forward. Climate change was real, the industry acknowledged. But instead of being a threat to the bottom line it presented a golden opportunity to develop new markets. These were to be encouraged by establishing modest carbon dioxide emission reduction targets and modest carbon pricing, and promoting growth through government subsidies for clean technologies.[49]

Greenwashing and individual responsibility

Industry's twenty-first century narrative positions itself as a leader in the fight to lower emissions. This new stance celebrates industry's projects and improvements, ranging from carbon capture and storage to natural gas, advocates climate-focused legislation that serves industry's interests, and promotes individual responsibility as the key to reducing emissions. British Petroleum (BP) changed its name to Beyond Petroleum in 2001 and inaugurated a major branding campaign to present itself "as the solution, rather than a contributor, to climate change."[50]

ExxonMobil's 2020 annual report emphasized how the company was "doing our part" to "responsibly develop new resources to ensure the world has the energy it needs while also minimizing environmental impacts."[51]

Fossil fuel companies are responding to pressure from lawsuits, shareholders, protests, public opinion, and voters. Their public reputation matters to them, as does the threat posed by growing support for measures that would interfere with profits. Their greenwashing aims to co-opt climate actions that could undermine the economic model they currently control.

The fossil fuel and associated industries, which openly participated in and helped to shape the Paris Agreement, opposed President Trump's withdrawal. "Climate change is real" intoned General Electric CEO Jeff Immelt, decrying Trump's action, and "industry must now lead and not depend on government." Tesla's Elon Musk announced, "Leaving Paris is not good for America." Exxon's CEO called the Paris Agreement "an effective framework for addressing the risks of climate change" and urged Trump to remain in it.[52]

Campaigns that encourage individuals to see their own consumption decisions as the cause of—and the solution to—climate change seek to divert public attention from industry responsibility. Mary Annaïse Heglar calls it "the narrative that has both driven and obstructed the climate change conversation for the past several decades. It tells us climate change could have been fixed if we had all just ordered less takeout, used fewer plastic bags, turned off some more lights, planted a few trees, or driven an electric car."[53]

"Know your carbon footprint!" a BP campaign urges, offering "a personalized set of results about you and your carbon emissions." After calculating emissions, the site offers tips on how to offset one's emissions, because "small actions can make a big difference."[54]

By emphasizing individual responsibility, though, Heglar continues: "We let the government and industries—the authors of said devastation—off the hook completely. This overemphasis on individual action shames people for their everyday activities, things they can barely avoid doing because of the fossil fuel-dependent system they were born into." Instead of trying to change the system, corporations urge, we should just change ourselves.[55]

In their publicity campaigns to individualize responses to climate change, fossil fuel companies turned to a playbook they developed in response to mid-century attempts to regulate waste caused by the expansion of industrial production, processing, and packaging. During the late 1950s, the National Association of Manufacturers and other industry groups mobilized to individualize the problem of waste. The American Can Company, the Owens-Illinois Glass Company, the Dixie Cup Company, and other throw-away products manufacturers organized the Keep America Beautiful (KAB) campaign. "The centerpiece of the organization's strategy was its great cultural invention: *litter*." The industries were not the problem: individuals were to blame for improper disposal of trash. As Heather Rogers explained in *Gone Today: The Hidden Life of Garbage*, "KAB paved the way in sowing confusion about the environmental impacts of mass production and consumption, today a favorite tool in the corporate greenwashing world."[56]

Industry (mining, agriculture, manufacturing, petrochemical production) produces seventy tons of waste for every ton of household waste. And most household waste is a result of decisions made by industries, like packaging policies and the proliferation of disposable products. But most of us have accepted the commonsense idea actively promoted by industry that it's *our* responsibility to solve the waste problem.[57] Environmental regulation, KAB warned, would only destroy jobs.

At the first Earth Day in 1970, speakers challenged capitalism and the modern system of production and consumption. On Earth Day 1971, KAB released its iconic "crying Indian" advertisement, intoning "People start pollution. People can stop it." In

1990, the twentieth-anniversary Earth Day celebration brought business and union leaders onto the board of directors and adopted the slogan "Who says you can't change the world?" The "corporate takeover of Earth Day" went hand in hand with "its heightened focus on individual environmental action."[58]

Consumers were thus primed to see individual behavior as the cause of environmental ills when fossil fuel companies jumped onto the bandwagon. Rather than driving particular forms of consumption, they claim, they are just responding to individuals' needs and demands. Individuals should focus on their own "carbon footprint" rather than challenging the fossil-dependent system.[59]

Fossil fuels and the economy

As we've discussed above, fossil fuels are woven into virtually every sector of our economy. So the fossil fuel industry is not alone in promoting the idea that our well-being depends on them.

"A ban on fracking in the United States would be catastrophic for our economy," the US Chamber of Commerce wrote in 2020. Such a ban would, it claimed, within five years, "eliminate 19 million jobs and reduce US Gross Domestic Product (GDP) by $7.1 trillion. Tax revenue at the local, state, and federal levels would decline by nearly a combined $1.9 trillion. Natural gas prices would leap by 324 percent, causing household energy bills to more than quadruple. By 2025, motorists would pay twice as much at the pump for gasoline as oil prices spike to $130 per barrel, while less domestic energy production would also mean less energy security."[60]

In the early days of the industrial economy, technology and fossil fuels mostly helped the rich get richer at the expense of the poor. In *Fossil Capital*, Andreas Malm argues that industrialists shifted from water power to fossil fuels in large part in order to relocate factories from remote mill towns on rivers to urban centers, with their large pools of labor, enabling greater control over their easily replaceable workers. Anyone who's read

Charles Dickens or Friedrich Engels has images of the "dark, satanic mills" that characterized the early Industrial Revolution.[61] Control of energy resources helps to define power and inequality in today's world still.

In the twentieth century, though, fossil fuels became the basis of consumer society, drawing workers into an "American way of life" based on "privatized social reproduction, single-family housing, and automobility." Oil "gave millions of Americans the wages, the public infrastructure, and the financial institutions to mobilize a whole host of energy-intensive machines in everyday life." Through the New Deal, according to historian Matthew Huber, American workers exchanged their radical dreams of power in the workplace for the consumer's dream of power—literally, based on fossil fuels—in private life.[62]

For President Franklin D. Roosevelt, cheap energy was to be the pillar of "the abundant life," and his New Deal promoted "uniquely wasteful patterns of mass oil (and other commodity) consumption." As one petroleum trade journal asked in 1950: "Did you know that a nation's progress (and its standard of living) can be measured pretty well by its consumption of petroleum?"[63] That this equation seems obvious and commonsensical to us today reflects just how successful the industry and its government sponsors have been in inculcating us with the idea that our individual well-being is inherently dependent on the well-being of fossil fuel producers.

When several US cities tried to hold fossil fuel companies accountable in court for damages caused by climate change, Chevron's vice president argued, "Reliable, affordable energy is not a public nuisance but a public necessity." The judge agreed that companies could not be blamed. "Our industrial revolution and the development of our modern world has literally been fueled by oil and coal," he wrote. "Without those fuels, virtually all of our monumental progress would have been impossible." "Would it really be fair to now ignore our own responsibility in the use of fossil fuels and place the blame for global warming on those who supplied what we demanded? Is it really fair, in light of those benefits, to say that the sale of fossil fuels was unreasonable?"[64]

The judge's statement that fossil fuel companies "supplied what we demanded" is somewhat disingenuous; it's more accurate to say that they flooded us with products, advertising, and insecurity that made us willing collaborators in the system. Yet he's correct to say that we can't really extricate the fossil fuel companies from the industrial economy in which they are imbedded. While fossil fuel companies have taken explicit steps to influence the debate on climate change, the deeper impact may stem from their profound imbrication in virtually every aspect of modern society.

WHAT KINDS OF POLICY SOLUTIONS DO ENVIRONMENTALISTS PROPOSE?

"Environmentalism" is a capacious category, encompassing people and organizations from across the political spectrum and with very different ideas about what caring about the environment means. US environmental organizations in the early twentieth century focused on preserving "wilderness" areas for the enjoyment of the privileged few, often by displacing Indigenous peoples who lived in the areas that colonizers wished to preserve. Many supported racist immigration and eugenics policies. Some are only now confronting their roots in white supremacy.[65] Meanwhile, environmental justice organizations in the United States and beyond are based in marginalized communities for whom environmental rights to land and water are inseparable from demands for social and racial justice, and for some, decolonization and a new international economic order.

Increasingly, established environmental groups like the Nature Conservancy, the Environmental Defense Fund, the World Wildlife Fund, Conservation International, and the Natural Resource Defense Council, sometimes dubbed the "big green" organizations, have moved into collaboration with polluting corporations to help improve their production practices. Newer, youth-led grassroots organizations like Extinction Rebellion and the Sunrise Movement have more radical goals, preferring direct action and civil disobedience to collaboration. They deride corporate alliances as greenwashing, legitimizing the actors and systems that are causing our environmental crisis.[66] Globally, Indigenous and peasant organizations struggle to defend their land, water, and forests from destruction in the name of economic development.

This division is reflected in the debate about the future of fossil fuels. The industry, mainstream environmental organizations, and many unions promote largely unproven carbon capture technologies. So do most world governments, and the IPCC, which argued for CCS as an essential component of a pathway to 1.5 degrees Celsius. Many grassroots organizations, though, reject CCS entirely and seek to end all new fossil fuel development and rapidly wind down existing production. The many organizations pushing for the global Fossil Fuel Non-Proliferation Treaty or calling on President Biden to #BuildBackFossilFree emphasize the disconnect between stated commitment to reducing emissions and actual and projected fossil fuel production.[67]

David Roberts proposed the term "climate hawks" to describe those of a diversity of political views who believe that the climate emergency transcends all other concerns.[68] Environmentalists may oppose wind, solar, hydropower, or nuclear projects because of their impact on land, forests, or wildlife, to say nothing of people; climate hawks say the nature of the emergency requires that we subordinate these issues and focus exclusively on lowering emissions.

The climate left argues, in contrast, the climate emergency is so embedded in our capitalist economic system that effective action on the climate must challenge capitalism itself.[69] Eco-socialists tend to advocate expansion of the social welfare state and government investment to build a new green—and fair—economy. Degrowth advocates offer an even more profound critique of industrial society: rich countries must significantly reduce their use of the planet's resources.

All of these positions and organizations may consider themselves environmentalist, but they encompass profound differences in perspective that lead them to advocate different types of policy approaches. They even differ on what's probably the most significant policy proposal to emerge in recent years—the Green New Deal.

WHAT IS THE GREEN NEW DEAL?

The Green New Deal refers either to a concept or to one of several different specific proposals that began circulating, primarily in the United States and Europe, in the late 2010s. The name references the New Deal of the 1930s, through which the federal government enacted a fundamental reordering of the US economy. During the Great Depression, a plethora of new government agencies, laws, and regulations were created to bring about a national recovery in agriculture and industry, including public investment in a social safety network, infrastructure, and jobs, and to enact social and economic protections for workers, farmers, and the population at large. Green New Deal proponents say today's climate crisis requires another profound transformation of our economy—and robust government investment and intervention to bring it about. It takes seriously the IPCC's warning that fundamental structural change is necessary to address climate change. And it recalls the New Deal's use of federal power to regulate and tax industry in the interest of economic redistribution and social protections: to place the common good above the profit motive, or at least to harmonize the two.

In the United States, the term Green New Deal usually refers to H.R. 109/S. 59, first introduced by Representative Alexandria Ocasio-Cortez and Senator Edward Markey in February 2019. After the 2018 midterm election that brought a Democrat majority and several progressive new voices into the House of Representatives, youth activists in the Sunrise Movement demanded that Congress take action on the climate. Sunrise drew less from the mainstream environmental movement and more on recent youth mobilization ranging from Occupy Wall Street to immigrant Dreamers to #BlackLivesMatter to fossil fuel divestment campaigns. Its approach to the climate crisis, unlike that of many established environmental organizations, grew from a social justice perspective.

That fall, two hundred Sunrise Movement youth occupied the office of House Minority Leader Nancy Pelosi to demand a Green New Deal. Their protests were followed by a January 2019 letter to Congress from a wide range of progressive organizations outlining their vision for climate justice. The letter called for an immediate end to fossil fuel leasing and subsidies, leading to a complete phasing out of fossil fuels in power generation, industry, and transportation. It called for immediate federal controls on CO_2 emissions under the provisions of the Clean Air Act. And it announced that signers would "vigorously oppose any legislation" that "promotes corporate schemes that place profits over community burdens and benefits, including market-based mechanisms and technology options such as carbon and emissions trading and offsets, carbon capture and storage, nuclear power, waste-to-energy and biomass energy."[70]

Many of the six hundred signers of the letter were small, local, or faith-based organizations. Large mainstream environmental groups like the Sierra Club, the Natural Resources Defense Council, and the Environmental Defense Fund were conspicuously missing from the signatories. Others, like Greenpeace, Friends of the Earth, 350.org, the Sunrise Movement, the Rainforest Action Network, the Indigenous Environmental Network, and Amazon Watch did, however, sign on. So did the Labor Network for Sustainability, an organization drawing together the left wing of the labor movement in support of climate-change action.

By February 2019, the newly elected representative Alexandria Ocasio-Cortez of New York had joined with veteran congressman Edward Markey of Massachusetts to propose H.R. 109. The resolution called for massive federal commitment to a ten-year action plan on climate change—"a new national, social, industrial, and economic mobilization on a scale not seen since World War II and the New Deal era"—aimed at reducing emissions to net zero. By linking investment in technology and infrastructure to socioeconomic transformation, jobs, and protection of frontline communities, it explicitly attempted to overcome historical divisions between the labor and environmental movements and between grassroots environmental justice and mainstream "big green" organizations.[71]

The Green New Deal (GND) proposal broke new ground by establishing an ambitious, enforceable target and embedding climate goals in socioeconomic transformation that prioritized the rights of workers and marginalized communities. But in terms of specific policy, the resolution left many issues open to future debate. In contrast to the demands in the January letter, the GND proposal called for "net-zero" emissions—meaning that it held open the door for continuing emissions and market-based policies like carbon trading and offsets. It called for meeting 100 percent of US energy needs with "clean, renewable, and zero-emission energy sources," implying that nuclear, biofuels, or even natural gas—options rejected by the January letter—could continue.

The right wing was quick to mock the proposal as unrealistic, fantastical, and anti-American. Fox News devoted more time to the topic than any other television news outlet.[72] Donald Trump ridiculed the GND for supposedly eliminating "Planes, Cars, Cows, Oil, Gas & the Military."[73]

Critics on the left also took issue with aspects of the GND proposal. It ignores the social and environmental costs of supposedly "clean" energy. Its "social democratic fixes" based on big increases in alternative energy would still rely on resource extraction from the Global South. Some called it "green colonialism" that would "be delivered by the very same entrenched economic interests, who have willingly sacrificed both people and the climate in the pursuit of profit. But this time, the mining giants and dirty energy companies will be waving the flag of climate emergency to justify the same deathly business model." Contrary to Trump's characterization, the proposal avoided confronting the role of the US military in contributing to climate change. Many on the left, though, saw the GND proposal as an opening to press for a "radical Green New Deal" that could inspire broad popular support with redistributive and social welfare programs and transform the US and global economic system.[74]

In the fall of 2020, then-candidate Biden confused some by variously applauding the GND as a "framework," stating that he did not support the GND, and offering his own "Plan for a Clean Energy Revolution and Environmental Justice."[75]

Once in office, Biden wasted no time in making climate a priority. He named John Kerry as his climate envoy, rejoined the Paris Agreement, banned new fracking leases on federal lands, and set a bold (though, to many, nowhere near bold enough) new target for reducing emissions: 50 percent below 2005 levels by 2030. While many celebrated these moves, critics said it was far from enough. The new target wasn't as bold as it might have seemed, since the Obama administration, when it signed the Paris Agreement in 2015, had already set a US target of a 28 percent reduction by 2025. Biden's announcement seemed to align with the IPCC recommendation that global emissions fall by that amount to even hope to keep warming to 1.5 degrees Celsius. But that's a global average; big emitters actually need to reduce far more.

Furthermore, as energy journalist David Roberts pointed out, climate politics has been rife with grand gestures, statements, and targets, which have thus far had little effect on reducing emissions. Until the Biden administration translates the goals into enforceable policy, his targets may have little practical effect.[76]

Even if the US fulfilled the emissions target of 50 percent below 2005 levels by 2030, it would remain the world's number-two culprit in annual emissions and number one in cumulative emissions, as well as close to the top in per capita emissions. The US

Climate Action Network calculated that, to do its "fair share," the United States would have to reduce its emissions by a shocking—but not impossible—195 percent by 2030.[77]

Biden's March 2021 American Jobs Plan was clearly influenced by GND goals and proposals. The plan put forward federal investment in electric vehicle production, public transit, research and development in "clean" energy (including nuclear, biofuel, and carbon capture technologies shunned by many environmentalists), building upgrades for energy efficiency, and significant guarantees for workers and affected and marginalized communities. Most GND proponents celebrated Biden's plan, though it relied heavily on incentives for the private sector, rather than the public sector expansion advocated by the original proposal. Direct public spending, explained Kate Aronoff, would be crucial to the GND vision of building public support for ambitious climate goals.[78]

The fossil fuel industry was quietly optimistic. Biden's plan would direct significant funds its way for cleaning up emissions and allow the industry to continue exporting its products. The petrochemical industry doubled down on its position that fossil fuels would solve global poverty. "Without plentiful, reliable and low-cost fossil fuels, the world would be in a very different place," an industry spokesperson insisted. "It would be less advanced, much less prosperous, have much shorter life expectancies."[79] But in fact the plan quietly assumed that the rest of the world would continue to provide the resources for the high levels of consumption in the United States.

Europeans countered with two very different road maps with confusingly similar names. The Democracy in Europe coalition, a coalition of progressive organizations and parties aimed at democratizing the EU, presented *A Blueprint for Europe's Just Transition* while the European Union offered a counterproposal in its European Green Deal.

The EU's Green Deal explicitly left out the term "new," distancing itself from the New Deal's commitment to an activist government promoting redistributive goals. Like H.R. 109, the EU proposal outlined goals in general terms and called for legislation to achieve them. But even as it shared just transition language with its US counterpart, the European document is clearly more growth, business, and market-oriented. Its reduction targets were less aggressive and longer term: halving emissions by 2030 and achieving net-zero by 2050. And it emphasized market-based measures, including expanding the European ETS and developing "new pricing instruments" to "ensure that the relative prices of different energy sources provide the right signals for energy efficiency." Regarding industry subsidies and exemptions, the proposal offered suggestions rather than mandates: "Fossil-fuel subsidies should end," and "the Commission will look closely at the current tax exemptions including for aviation and maritime fuels and at how best to close any loopholes." Beyond the EU, policies "may include ending global fossil fuel subsidies . . . phasing-out financing by multilateral institutions of fossil fuel infrastructure, strengthening sustainable financing, phasing out all new coal plant construction, and action to reduce methane emissions." Or, they may not!

Europe's reformed energy sector would phase out the use of coal and be "based largely on renewable sources" while continuing the use of natural gas and increasing reliance on carbon capture and storage. While the proposal gave a nod to the issue of natural resource overuse and the need for recycling and a "circular economy," it relied on "measures to encourage businesses" to move in this direction rather than regulation. It insisted—ominously, from a Third World perspective—that "access to resources is also a strategic security question for Europe's ambition to deliver the Green Deal."[80]

Democracy in Europe Movement's blueprint, the Green New Deal for Europe (GNDE), offered quite a different vision, and much greater detail. The GNDE adopted the Paris Agreement's 1.5-degrees Celsius warming target, aimed to achieve net-zero emissions by 2025, and called for economic and ecological reorganization away from a profit-and-growth economic system to one that prioritizes human needs and respects planetary boundaries. The GNDE's *A Blueprint for Europe's Just Transition* included detailed policy proposals for radical transformation in every sector of the economy including Green Public Works and a Europe-wide regulatory and legal framework for a just transition. It offered paths to remake infrastructure, industry, agriculture, energy, and transportation, to reduce or eliminate military spending and war, to democratize the EU with grassroots people's assemblies to press elected leaders to action, and to challenge growth-oriented capitalism.[81] Its advisory board included European Green, Socialist, and Communist Party veterans, as well as Jane Sanders (Bernie's wife) and Naomi Klein.

Just one example from the GNDE Blueprint: it called for an "energy allowance" for every household granting free electricity sufficient for basic cooking and heating needs. Beyond the allowance, prices would rise steeply. The allowance addresses social justice goals by ensuring basic security. The system is redistributive because the rates rise with overuse, essentially taxing the rich at higher rates. It encourages conservation because it makes overconsumption progressively more expensive. And it creates revenues for other aspects of a just transition.

The various proposals are important because they are being promoted in some of the planet's most polluting countries. But clearly the Green New Deal (GND) concept covers a wide range of potential policies. Some proposals rely on business-friendly, market-based incentives, while others directly challenge the fossil fuel industry and the whole fossil economy. Some aim at deep socioeconomic transformation. The strongest ones address the historical responsibility of the Global North, offer enforceable targets, and join with movements for economic transformation away from fossil capitalism. Such coalitions and mobilization will be crucial in order to ensure that the more powerful—and most responsible for emissions—can be forced to comply with the spectrum of measures so urgently needed.

CONCLUSION

Scientists and activists agree that a broad spectrum of policy approaches is necessary to confront the climate crisis. There is no single magic bullet, given the multiple sectors that contribute to emissions. The IPCC is very clear, given the severity of our current crisis: "No single option is sufficient by itself. Effective implementation depends on policies and cooperation at all scales and can be enhanced through integrated responses that link mitigation and adaptation with other societal objectives." The panel called for "rapid and far-reaching transitions in energy, land, urban and infrastructure (including transport and buildings), and industrial systems. These systems transitions are unprecedented in terms of scale . . . and imply deep emissions reductions in all sectors, a wide portfolio of mitigation options and a significant upscaling of investments in those options."[82]

Energy journalist David Roberts put it even more bluntly: "The actions necessary to hold to 2 degrees, much less 1.5 degrees, are simply outside the bounds of conventional politics in most countries. Anyone who proposed them would sound crazy." It's not enough just to halt new development of fossil fuel capacity or even just to eliminate all use of coal. Just using up what remains in currently operating oil and gas fields would bring us above the 1.5 degrees Celsius goal.[83]

In their "wide portfolio" of potential actions, the various paths proposed by the IPCC all rely on unproven carbon capture technologies, to enable production and economic growth to continue. Ecological economists point out that the IPCC ignores a much simpler and more obvious alternative: that we find ways to slow or reverse economic growth in the overconsuming countries. "The principle of reducing energy and resource use represents a safer and more ecologically coherent approach to climate mitigation," conclude Jason Hickel and his coauthors.[84] But, thus far, most policy proposals have studiously ignored this option.

Given how long we have known the facts about climate change and its horrific impacts on human survival on the planet, why haven't we been able to implement more effective policies to change course? Some factors specific to US culture, economy, and politics place this country at the extreme end of failure to restrain emissions. But in fact no industrialized country has successfully done so.

Part of the reason is that virtually every sector of industrial society depends on fossil fuels. A huge array of interests is invested in the status quo. Industries lobby vociferously for their interests, while voters are reluctant to support policies that imply sacrifice. The GND has tried to reframe the debate with proposals that show how an emissions-free economy could actually be better for the majority. But many entrenched interests staunchly oppose the kind of substantive, redistributive change the GND proposes, and most versions of the GND don't confront the outsized impact on the planet's resources that even an emission-free and more equal Global North would continue to have.

CHAPTER 3

WHAT CAN I DO AS AN INDIVIDUAL?

Given the paralysis of our institutions in confronting the climate emergency and the fact that those of us in the United States contribute an outsized proportion of the planet's emissions, it's natural to think about what kinds of actions we can take individually to change things. We are all cogs in the fossil fuel economy—it depends on us to keep consuming. What if we refuse to participate? What if we identify how we are complicit in the system and withdraw our consent?

In a sense, any action we take falls under what we can do "as an individual." But our dominant ideology encourages us to think of ourselves primarily as consumers and to seek individual solutions through new products like energy-efficient appliances or cars, locally grown foods, or lifestyle changes like eating less meat or reducing air travel.

A study of government recommendations and science textbooks found they focused most commonly on recycling and on buying energy-efficient products from lightbulbs to cars. This approach suggests that we can save the planet through minor consumer choices, by choosing one more energy-efficient product over another.

Manufacturers of virtually every product now advertise their green credentials or offer green products. But usually these green alternatives actually encourage more consumption: that's why companies manufacture them.

Obviously, we have to consume some things to survive—food, shelter, clothing, and medicine. But every form of consumption has environmental costs. Green products still use resources and create waste, including emissions. If a new green product adds to overall consumption, it's contradicting its "green" label.

The highest-impact changes we could make in our lifestyles aren't about choosing one product over another. Only a very few of us consider the climate implications of what may be by far the most significant individual decision we make: whether to have children. (Some argue this notion stretches the concept of individual responsibility excessively, since it holds individuals responsible for other individuals' emissions.) After the question of having children, the highest-impact changes an ordinary high-consuming individual (that is, almost all individuals in the United States), can make are decisions *not* to consume: going car-free, avoiding a transatlantic flight, or reducing or eliminating meat from our diets. The choice *not* to buy something, or to consume less, is always the most climate-friendly choice.[1]

I'll look at some popular green products and some other individual actions here, avoiding popular but small-scale decisions like buying energy-conserving lightbulbs and recycling and focusing on the biggest emitters: transportation and food. But in the end, I'll argue in order for our individual actions to be politically meaningful, we have to go beyond thinking of our actions as personal lifestyle and consumption adjustments. We can't buy our way out of the problems our economic system has created. It's through organized social movements and political action that we can create the kind of pressure necessary for real structural change. Individuals can build and participate in these movements. We can create alternative, low-carbon forms of social organization. We can link local campaigns and actions to larger goals. We can challenge the structures that keep us on the fossil treadmill. We can recognize the urgency of the climate crisis and demand change. We can employ tactics like mobilization, direct action, and civil disobedience, which pressure our institutions to take the kind of action we need.

This doesn't mean that individual and lifestyle choices are meaningless. Recognizing how the fossil economy works and our own role in it can help us to imagine and fight for larger political, social, and economic change. But it's crucial we take that next step and not limit our political imagination to consumer choice.

SHOULD I BUY A PRIUS?

No! Here's why. First, let's look at how hybrid and electric cars work and how they're manufactured. Then, we'll address the bigger issues surrounding individual cars.

Conventional cars rely on internal combustion engines—you fill the tank with gasoline that the engine burns directly. Fully electric cars use no gasoline—you charge a large battery by plugging into the electrical grid, and the car operates until the battery runs down. Instead of filling the tank, you plug it in again to recharge. Hybrid cars use different combinations of these technologies, with an internal combustion engine and a battery that charges by plugging into the grid or by capturing and storing some of the energy the car produces (e.g., when you brake). So, electric cars use no gasoline at all, and hybrids use a lot less than conventional cars; both types greatly reduce tailpipe emissions. As hybrid and electric cars become more common and the technology advances, their high cost is also coming down.

But that's not all there is to the story. Every car requires energy (generally from fossil fuels) and other resources (metals, chemicals) to produce. More emissions are created in the manufacturing process for more technologically advanced hybrid and electric cars than for traditional gasoline-powered cars. (It depends in part on where they are manufactured and what electricity source is used.) Still, their lower or nonuse of fossil fuels when operated mean that over their life cycle, their emissions are lower overall.[2]

Hybrid and electric cars also need batteries to store energy. These batteries rely on the same nonrenewable resources used for storage systems in "renewable" electric grids, including lithium, cobalt, manganese, and graphite, all elements whose extraction entails significant environmental and social costs, often in the Third World.[3]

If a car is wholly or partly electric, it relies on the existing power grid to charge the battery. So unless the grid is emission-free, you still create emissions when you recharge. Even with a fossil-powered grid, though, it's more efficient, and creates fewer emissions, to use electricity from the grid than to burn gasoline in your car. A hybrid or electric car can travel 50 to 80 miles on the equivalent of a gallon of gasoline, something no conventional car can currently achieve.

A potential downside of this efficiency is that better gas mileage frequently encourages people to drive *more*. Driving more or maybe buying a bigger car because you can use less gas is one example of what's called the rebound effect, the fact that increased energy efficiency doesn't generally result in emissions reductions. Instead, increased efficiency frequently encourages *more* energy use. As products become more efficient, people may buy bigger ones, replace them more often, use them more often, or buy new things with the money they save. Since the profitability of companies depends on people consuming more, they flood us with new opportunities, new products, and new expectations, using efficiency as an added lure.[4]

Of course, we could avoid the rebound effect. Instead of driving or consuming more, individuals could combine efficiency with reduced consumption and a shift to low-emission activities. But that's not what usually happens, especially at a societal level. There are too many structural factors pushing us to consume more and better products—including more and better cars.

Hybrid and electric vehicles could lead us to the dystopian future imagined by Thea Riofrancos: "A world buzzing with hundreds of millions of Teslas (or worse, e-Escalades), made with materials rapaciously extracted without the consent of local communities, manufactured under a repressive labor regime in polluting factories—in other words, a world not unlike our own, but powered by wind and sun."[5]

If this is not the future we desire, we need to rethink our transportation problem, like the other components of our climate emergency, beyond technical fixes, and certainly beyond individual consumer decisions. Whether it runs on gasoline, electricity, a combination, or something else entirely, two tons of metal to transport a one-hundred- to two-hundred-pound person from one place to another just doesn't make sense. Resource extraction, automobile manufacture, roads, parking infrastructure (land use, building, and maintenance), traffic fatalities, and end-of-life disposal all constitute different costs of our reliance on personal automobiles.[6] Is it possible to imagine a better world, with fewer cars?

Worldwide, we humans buy as many as eighty million new cars every year—but that's not evenly distributed. In the US, 88 percent of households own a car. In the world's poorest countries, it's under 10 percent. And in the two most populous countries, China and India, it's 17 percent and 6 percent.[7]

Lack of mobility is a serious problem for the majority of people. We should not acquiesce to a world that apportions mobility so unequally. But one that multiplies cars—and emissions—enough to generalize the US pattern is also untenable.

Even among wealthy countries, there are vast differences in how much people drive. In the United States, mobility is extraordinarily difficult without a car. Here, each of us drives around 13,000 miles a year. In Europe, it's half that. In Japan, it's even less. It's not poverty that keeps people in those countries from driving; it's the built environment and public transportation infrastructure that make cars less necessary and less attractive.[8]

It's fine to choose a car with good gas mileage. But that shouldn't replace fighting for an infrastructure that enables ordinary people in countries like Japan or Denmark to rely on bikes and public transportation, despite inclement weather, the need to transport one's children, and other common objections from people in the United States. As former Bogotá mayor and livable streets advocate Enrique Peñalosa said, "An advanced city is not one where even the poor use cars, but rather one in which even the rich use public transport."[9]

In the United States, government and industry have collaborated to create urban, suburban, and rural sprawl, a capacious interstate highway system, and automobile production, while turning their backs on public transit and high-speed rail. These decisions have made us much more car-dependent than other industrialized countries. They've also taught us to associate our individual vehicles with freedom. These realities will not be easy to undo. But they are exactly what we must challenge in order to start the restructuring we need. If buying a Prius, or changing your own transportation habits, can move you towards becoming a transportation activist, then it's a step in the right direction. But if you buy a Prius and think you've done your part, you're just greenwashing an unsustainable, car-centric system.

Focusing on individual responsibility downplays what shapes our individual choices. If public transportation is plentiful, reliable, and cheap, while gasoline, tolls, and parking policies make driving inconvenient and expensive, more people are going to use public transit. If, as in most of the United States, public transportation is expensive, inconvenient, and unreliable, while the built environment, roads, gasoline, and parking policies make driving essential, cheap, and convenient, people are more likely to rely on individual cars. Individuals make choices, but larger structures determine what choices are available to us and shape our values and expectations.

WHAT ABOUT GIVING UP MY CAR AND USING UBER AND LYFT?

New so-called ride-sharing services like Uber and Lyft are often promoted as cutting-edge innovations that use technology and collaboration to create a new, greener sharing economy. Indeed, Lyft started out as a kind of carpool system.

But what they've evolved into is a big moneymaking business that keeps more cars on the road, exploits workers, and undermines public transportation. The term "ride-sharing" is really a misnomer, as there is nothing "shared" in the system. Instead, big companies contract private drivers to sell chauffeur services to individuals, with the companies taking a cut. It's ride-selling, not ride-sharing.[10] While the term "ride-sharing" still dominates in the media, many scholarly studies use the term transportation network companies (TNCs). I'll adopt that terminology here.

When one of my oldest friends and colleagues, Robert F. Young, passed away unexpectedly in 2018, I attended his memorial in Austin, Texas, where he had been living and teaching. In some ways, I could say it was Rob who introduced me to environmentalism: one of our first conversations was in the back of an old VW van heading to California's Diablo Canyon to protest PG&E's nuclear power plant there in the late 1970s. Rob's love of the natural world and his deep commitment to both innovative analysis and direct action influenced me profoundly.

At the memorial, several speakers mentioned Rob's signature "Green Cities" course that he taught at Cornell for many years. A young man who introduced himself as John Zimmer talked about how much the course had influenced him—so much so that when he graduated, he went on to found Lyft!

I was surprised, because by 2018 my impression of Lyft was that it was just another wasteful, exploitative embodiment of the new gig economy.

But it turns out that it began on Rob's Cornell campus when Zimmer began Zimride, a platform to facilitate carpooling, to get people out of their individual cars. Zimride mobilized social networks to help students find classmates who were traveling similar routes, allowing them to share rides and expenses. Although Zimmer explained that he developed the platform because "carpooling sucks," Zimride was in fact a ride-sharing, carpooling system.

Its 2012 reincarnation as Lyft was something different. Although billed as a way to provide an alternative to car ownership, Lyft, like Uber, ended up adding cars, and car trips, to the roads. These companies make car travel more convenient and accessible, while covering it with the cachet of being green, trendy, and part of the new sharing economy.

Advocates claim that Uber and Lyft are more convenient, cheaper, and more flexible than taxis. Indeed, one of the casualties of the TNC industry has been the taxi industry. But we should be careful about assuming "more convenient, cheaper, and more flexible" is necessarily a good thing. Free market competition can foster something good for some consumers but not necessarily for workers, poor people, and the environment.

The taxi industry is highly regulated, which means that public policy sets rules and limits. This made taxi driving a secure path to financial stability for many drivers. TNCs jumped into a completely unregulated market and have taken full advantage of it. The European Union and several European and other localities have attempted to regulate or ban TNCs in ongoing legal battles.

Flooding the roads with cars offering cheap rides, TNCs aggressively underpriced taxis in order to capture the market. The *American Prospect* compared their tactics to those of Amazon and Walmart. Backed by big subsidies from venture capitalists, these companies can sell below cost to eliminate competition. But as in the case of Walmart, there is a "high cost of a low price."[11]

TNC workers, like others in the so-called gig economy, operate almost completely outside the scope of protective labor legislation. (Companies like Walmart maneuver to avoid labor protections by keeping workers part-time.) Taxi drivers, some of whom invested life savings or went into debt for a coveted taxi medallion, saw their customer base shrivel. A cluster of suicides among New York City taxi drivers in 2018 tragically highlighted this impact.[12]

There are racial implications to the shift also. TNC workers and users tend to be whiter and more educated than those who drive and use taxis. Immigrant and minority communities, writes Steven Hill, are "triply aced out by this techno-transport 'disruption,' i.e., losing taxi service as it declines vis-à-vis Uber; losing a once-viable occupation for low-skilled, English-limited immigrants; and sometimes having their neighborhoods redlined by ridesharing services."[13]

Meanwhile, by making single-occupancy transit cheaper and more convenient, services like Uber and Lyft have increased the number of cars on the road and miles driven everywhere they operate. They increase CO_2 emissions directly, by adding to the miles driven, and indirectly, by adding to traffic congestion. TNCs can undermine policies that cities have tried to implement to reduce traffic, like raising parking fees or implementing road space rationing (like allowing only even- or odd-numbered license plates to enter the city on certain days). TNC users can enjoy the benefits of driving without having to park or follow rules.

In 2019, UCLA students were taking eleven thousand ride-sourced trips a week just to get around campus. Users generally report that TNCs replaced walking, biking, or using public transit. In San Francisco, Uber and Lyft were responsible for half of the city's 60 percent increase in traffic congestion between 2010 and 2016. In the nine largest US metro areas, they added 5.7 billion miles of driving annually, even as car ownership continued to grow more rapidly than the population. That's because many TNC users don't actually give up their cars—they keep them but use TNCs when parking or other regulation makes driving less convenient.[14]

In addition to adding more cars to the road and increasing the miles driven, TNCs undermine public transportation in the dense metro areas where they mostly operate. As ride-sourcing offers an alternative to wealthier passengers, they tend to abandon public transit. As ridership decreases, so do revenues, and services are then cut, leading to a downward spiral that makes public transit an ever less-viable alternative.

Transit systems rely on revenues from high-use lines in dense areas to subsidize lesser-used off-peak and more remote routes. So declining urban use has a disproportionate effect on the economics of the system. Furthermore, as the wealthy turn to private alternatives, they become less interested in contributing to or politically supporting the public sector. Overall, the wealthy have more political clout. TNCs become just another example of how those better off can insulate themselves from declining public services and hasten their decline.[15]

Public transit creates fewer emissions and more equality. Private car services do the opposite. The transit situation reflects some of the pervasive ways a market system can undermine poor people and the climate. The market rewards increased production for those who can pay for it. Without strong public action, there is simply no way the market is going to do the two things we most need: reduce our use of resources and redistribute the benefits of industrial society to prioritize the urgent needs of the poor over increased consumption by the rich. As transportation expert Bruce Schaller concludes, we should not be distracted by the false association of TNCs with a "sharing economy" and with the cachet of instantaneous apps. There's an old-fashioned form of ride-sharing that requires political will rather than venture capital and innovation: "frequent, reliable, safe, and comfortable public transportation."[16]

SHOULD WE ALL STOP FLYING?

Air travel is one of the most concentrated sources of emissions, and one of the most unequally distributed. And it's growing fast.

Emissions from flying constitute only about 3 percent of global emissions. But those emissions come from a very few people. Only 5 to 10 percent of the world population flies in a given year, and only 2 to 3 percent—the world's global elite—fly internationally on a regular basis. And a single international flight can create the equivalent, per traveler, of half of the global average of annual per capita emissions. So it makes sense for individual fliers to place changing their flying habits high among their climate priorities. For a member of the global flying elite, cutting out a single transatlantic flight a year is one of the most effective ways to reduce emissions, exceeded only by having one fewer child or living car-free.[17]

But things are going in the opposite direction, and we need more than individual decisions to change course. The number of flights taken has increased every year since 2009, except for a drop in 2020. It reached 3.7 billion in 2016 and is predicted to reach 7.2 billion by 2035.[18]

What impact can individual decisions have on that trajectory? In Sweden, the *flygskam*, or "flight shame," movement encourages individuals to change their flying habits. It gained added attention when Greta Thunberg traveled by boat to New York to speak at the UN's Climate Action Summit 2019. Twenty-three percent of Swedes reported reducing their air travel that year due to environmental concerns, although Swedish airports reported only a 5 percent reduction in traffic.[19]

Structural factors are also crucial. Sweden's generous vacation policy and strong rail system make slower travel a viable alternative for domestic travel. Policies that keep air travel cheap—including government subsidies and exemptions of jet fuel and international flights from national and international climate regulation and pricing—make flying more appealing. Business models like that of Amazon, based on rapid delivery, are increasing cargo flights. Most business, structural, and institutional factors are pushing us towards more, not less, flying.

We should also consider for a moment social justice questions. Less than one-fifth of the world population has *ever* been on a plane.[20] Do the rich have more of a right to air travel than the other four-fifths of the world? Carbon pricing measures that make flying more expensive could shift the trend of ever-increasing air travel, but would also exacerbate the nature of flying as a privilege for the super wealthy. There's no easy answer for how to ration something that is currently as coveted, harmful, and as unjustly distributed as flight. But cultural shift and regulation can place limits and make it more expensive, while public investment in alternatives can make other forms of travel more accessible at less environmental cost. One thing is clear: in a just and sustainable world, air travel is a luxury that must be significantly reduced.

SHOULD WE ALL BE VEGETARIANS?

Although we don't see the fossil fuels embedded in what we eat, emissions occur in every step of the process of food production, from land use change as forests are cut down for plantations, farms, and pasture to pesticides, fertilizers, greenhouses (which are anything but green), and farm machinery used in growing food to refrigeration and transportation of products (sometimes by air) from farm to factory (where it's processed and packaged) to supermarkets.

The way the US EPA calculates it, agriculture accounts for about 10 percent of US emissions. Agriculture and land use changes (i.e., deforestation) together account for about a quarter of global emissions. A life-cycle analysis that adds in inputs like fertilizers, pesticides, food processing, and transportation tallies agriculture's share of global emissions at up to 40 percent.[21]

Agriculture presses on other, interrelated planetary boundaries as well: it accounts for 50 percent of global habitable land use, 70 percent of global freshwater use, and 78 percent of global water pollution. And 94 percent of mammal biomass on the planet (excluding humans) now consists of livestock. The spread of agriculture is the prime threat to twenty-four thousand of the world's twenty-eight thousand endangered species.[22]

The role of meat

One agricultural culprit stands out above all others in both emissions and other environmental impacts: meat. Livestock—that is, animals raised for their meat, or for products like their milk and eggs—occupy 77 percent of the planet's farmland. If you use a life-cycle approach that includes the emissions from the animals themselves, from deforestation for pasture, and from crops produced to feed livestock, animal agriculture accounts for one half to three-quarters of the world's food production-related emissions and 18 percent of total GHG emissions. The pound of beef you might eat over the course of a few days creates more emissions than a gallon of gas that you burn in your car.[23]

Livestock production uses land directly and indirectly (for producing feed) and its expansion is a major cause of deforestation. Ruminants like cows emit 37 percent of anthropogenic methane, and their waste (manure and urine) is responsible for the majority of the world's ammonia and nitrous oxide emissions.

In the United States, some one-third of the land is used for pasture for livestock. Agriculture takes up another one-fifth, or 392 million acres, only 77 million of which produces food that we eat. The largest portion of agricultural land, 127 million acres, is dedicating to producing—you guessed it—feed for livestock. Another 84 million acres is used to grow grain for export, much of which feeds livestock abroad. (The remaining land is fallow at any given time or dedicated to nonfood use.) In total, 41 percent of the country's land is devoted to supporting livestock in one way or another.[24]

From an environmental, land use, or emissions perspective, growing grains to feed cows is an extraordinarily inefficient process. It takes 16 pounds of grain to produce one pound of meat, as well as 2,500 gallons of water, and 20,000 calories in fossil fuel. Globally, one-third of all grain, and an even larger proportion of oilseeds, go to feed livestock rather than people.[25]

In terms of emissions, "grains, fruits, and vegetables have the lowest environmental effects per serving and meat from ruminants the highest effect per serving. . . . Plant-based foods cause fewer environmental effects per unit of weight, per serving, per unit of energy, or per protein weight than does animal source foods across various environmental indicators," concludes an exhaustive study in *The Lancet*. Beef production (the worst culprit) emits 17 times more CO_2e than does that of grains, fruits, and vegetables. In fact, concludes *The Lancet* study, "[c]hanges in food production practices could reduce agricultural greenhouse-gas emissions in 2050 by about 10%, whereas a shift to plant-based diets could reduce emissions by up to 80%."[26]

Food for profit, not for people

Why has our global agricultural system chosen such inefficient use of our land and resources? The short answer is profit. It's more profitable to produce for rich people than for poor people. Farmers have sold their food in markets for millennia. Small subsistence farmers can, in theory, sell their surpluses and purchase items that they don't produce for themselves. But the rich and powerful have found ways to turn food production into a form of extraction: enclosing, taxing, colonizing, dispossessing, indebting, and enslaving subsistence farmers. That long historical process has brought us to today's agro-industrial system in which giant corporations control a fossil-dependent system that dispossesses small farmers, exploits workers, eats up government subsidies, creates outsized emissions and toxic waste, overuses pesticides, fertilizers, and antibiotics, and produces and sells ever-more-processed foods and soft drinks. The world is overloaded with food—the supply of food calories per capita has increased by more than a third since 1961—but while 2 billion of the world's adults are overweight or obese, 821 million people are undernourished, close to 1 billion suffer from food insecurity, and 150 million from crisis-level hunger.[27]

In the last century, fossil fuels, technological advances, and government policies enabled profit-minded corporations to take over small farms and rely on inputs like tractors, harvesters, and new fertilizers and pesticides to increase productivity. Every intensification pressed further on planetary boundaries, including GHG emissions, especially as petroleum-based chemical use became widespread.

The growth of agribusiness undermines small-scale farmers, who still produce 70 percent of the food consumed in poor countries yet also comprise most of the world's one billion hungry people. Poor farmers are losing their land, lack access to government support or protection, and are only harmed by food gluts, whether through domestic industrial agriculture, cheap imports, or food aid. International development specialist Timothy Wise explains: "Increasing the industrial production of agricultural commodities does almost nothing for the hungry. It may lower urban food prices slightly, if agribusiness monopolies actually pass the savings on to consumers. But within developing countries capital-intensive farming reduces employment in rural areas, which increases migration to urban centers and reduces wages for low-skill workers. For hungry farmers, those commodities that 'we' export to 'feed the world' can even make them hungrier, as cheap imports undercut local food producers."[28]

Corporate agriculture, meat, and hunger

Meat production offers a prime example of how emissions, inequality, and hunger are intertwined in the industrial agricultural system. What's most profitable is also what causes the highest emissions: selling grains to livestock producers, who then sell meat to the rich, while the poor go hungry. Our system, and our large government subsidies, reward big farms and increased production, not sustainability or justice. Like other kinds of luxury production and consumption, meat brings profit while contributing to more inequality and more emissions.

"Farmers must constantly increase production," Frances Moore Lappé explains in her 1971 classic *Diet for a Small Planet*, "regardless of the ecological consequences. And they must constantly seek new markets to absorb their increasing production. But since hungry people in both the United States and the third world have no money to buy this grain, what can be done with it?" The answer, she shows, is meat. Our system "feed[s] about 200 million tons of grain, soybean products, and other feeds to domestic livestock each year." US grain also feeds meat production abroad. "While most Americans believe our grain exports 'feed a hungry world,' *two-thirds* of our agricultural exports actually go to livestock—and the hungry abroad cannot afford meat."[29]

Hunger in the modern world, then, is caused not by an absolute lack of food but by lack of access. Poor people can't compete in the market without money. Even some countries undergoing severe famines continue to export food. "The problem is not a scarcity of land or food," Lappé reiterates in 1991, in the twentieth-anniversary edition of the book, "it is a scarcity of democracy."[30]

The world's former colonial and settler colonial powers are the biggest meat consumers, with Europeans eating about 80 kilograms of meat a year per capita, and Australians and US Americans eating over 100 kilograms.[31] The Third World suffers the worst social and environmental consequences of this global system.

Agriculture and climate refugees

The changing climate further narrows the window of survival for poor farmers, many of whom have already been pushed onto marginal lands. In a vicious cycle, emission-heavy agriculture contributes to climate change, which contributes to rising sea levels, drought, desertification, hunger, war, and ongoing deforestation, which then turn peasants around the world into climate refugees, desperately seeking new land to cultivate or a better life in the city or in a new country.

"For most of human history, people have lived within a surprisingly narrow range of temperatures, in the places where the climate supported abundant food production," wrote a major 2020 investigation. "As the planet warms, those regions are shifting. Entire nations will lose their ability to farm grains and vegetables. Faced with starvation, those who can leave will have little other choice." Potentially one out of every three of the planet's inhabitants will be living in lands that can no longer sustain human survival by 2070.[32]

Ongoing climate change, then, makes our unjust and unsustainable system even more unjust and unsustainable. One disturbing result has been the increased militarization of the world's borders, as wealthier (and generally more temperate) countries barricade themselves against the waves of climate refugees seeking to escape lands rendered uninhabitable or uncultivable by the actions of the very countries that are now refusing them entry. "Border enforcement is not only growing," writes Todd Miller in *Storming the Wall*, "but it is increasingly connected to the displacement caused by a world of fire, wind, rain, and drought."[33]

An alternative path: Agroecology and food sovereignty

So back to our original question: Should we become vegetarians? Reducing or eliminating meat, or animal product, consumption is one of the easiest and most effective ways of reducing your individual carbon footprint. The IPCC emphasizes the importance of "changes towards less resource-intensive diets."[34] But becoming a vegetarian won't change the system.

With diet, as with transportation, personal purification is not in and of itself a very effective form of political activism. But if changing individual behavior is a step into other forms of activism, then it can be part of making a difference. If it's just a way of withdrawing from the world and focusing on self-improvement, then the minuscule impact is not going to register in the bigger picture.

There are ways to reorient our global agricultural system into one that puts human needs above profits, takes power out of the hands of wealthy consumers and corporations and puts it into the hands of democratic institutions, and prioritizes both reducing environmental destruction and filling the basic needs of the global poor. Advances in science and technology could be used to foster sustainability and justice, if they were more democratically controlled. Peasant organizations around the world, even the UN's

Food and Agriculture Organization (FAO), have crystallized their vision of a socially and environmentally just agricultural system around two themes: agroecology and food sovereignty.

Agroecology minimizes fossil-based inputs and fosters diversity rather than monoculture, local rather than corporate control, long-term health of the land rather than maximum extraction, and social justice rather than profit. The FAO defines agroecology as "an integrated approach that simultaneously applies ecological and social concepts and principles to the design and management of food and agricultural systems. It seeks to optimize the interactions between plants, animals, humans and the environment while taking into consideration the social aspects that need to be addressed for a sustainable and fair food system."[35]

The international peasant organization Via Campesina explains that industrial agribusiness is "carried out by a small set of increasingly large corporations seeking to expand private profits." In contrast, "peasant agroecological farming" is "practiced by peasants and other small-scale food producers" who "seek to meet human needs by working with nature."[36]

Advocates say that despite the expansion of industrial agriculture and the pressures on small farmers on every one of the world's continents, peasant farmers still produce up to 80 percent of the food in the nonindustrialized countries, while occupying less than a quarter of the world's farmland. Even in middle-income countries of Latin America and Eastern Europe, small farmers produce over half of the food.[37]

Food sovereignty prioritizes the rights and needs of these small farmers, strengthening land rights, and reorienting government programs like agricultural credit and technical support to small farmers producing for domestic consumption rather than for livestock or for export. Peasant organizations from around the world defined the concept in the 2007 Nyéléni Declaration as "the right of peoples to healthy and culturally appropriate food produced through ecologically sound and sustainable methods, and their right to define their own food and agriculture systems.

It puts those who produce, distribute and consume food at the heart of food systems and policies rather than the demands of markets and corporations. It defends the interests and inclusion of the next generation. It offers a strategy to resist and dismantle the current corporate trade and food regime, and directions for food, farming, pastoral and fisheries systems determined by local producers." Food sovereignty, then, means democratic control over food systems.[38]

Closer to home, the City University of New York Urban Food Policy Institute outlined how a food sovereignty and agroecology approach could be incorporated into a US Green New Deal. Its seven-point approach calls for shifting public money spent on food, like SNAP (Supplemental Nutrition Assistance Program) and school lunches, to support agroecological farming and food workers' rights, reducing subsidies for ultra-processed foods, providing living wage jobs in food and agricultural sectors, shifting agricultural subsidies from agribusiness to small and sustainable farms, and regulating food waste by businesses and producers.[39] The concrete proposals remind us that our own government and its policies play an outsized role in the global agricultural system and that practical steps to change direction are within reach.

A just transition in agriculture would honor the rural livelihoods of small farmers on every continent, while acknowledging that our current agro-industrial system must be transformed to prioritize the rights of humans and nature. In the process, the world's overconsuming carnivores will need to greatly reduce their meat consumption. Changing personal consumption choices is a first step; the real impact comes when we transform the system.

WHAT ARE STRENGTHS AND WEAKNESSES OF PIPELINE PROTESTS AS A STRATEGY? WHAT ABOUT DIVESTMENT FROM FOSSIL FUELS?

Fossil fuels are the basis of our industrial economy and the main source of the emissions that are warming the planet. And fossil fuel companies and their allies throughout the business and manufacturing world are deeply committed to keeping our economic system of production and consumption growing. So far, they have largely succeeded in keeping the climate change debate focused on technological, market-based, and consumer-oriented solutions that avoid confronting the roots of the problem.

The urgency of climate change requires that keeping the remaining fossil fuels in the ground be the top priority. Organizations like Oil Change International emphasize that "to live up to the goals set forth by the Paris Agreement and to safeguard our climate for this and future generations, fossil fuel production must enter a managed decline immediately." New fossil fuel projects must be abandoned, and current projects retired expeditiously. This position is supported by scientific studies showing that a third of known oil reserves, half of gas reserves, and 80 percent of coal reserves must remain untapped even to remain within the upper limit of two degrees Celsius. Existing and planned energy infrastructure alone would exceed the 1.5 degrees Celsius goal.[40]

Increasingly, climate activists worldwide insist on confronting the fossil fuel industry directly. Effective climate action means stopping the extraction and burning of fossil fuels. That requires we undercut the power of industry to shape legislation and regulation to its advantage.

Pipeline protests and divestment campaigns place ending fossil fuel use at the center of climate activism. Protests against specific fossil fuel companies and fossil fuel projects and infrastructure form the core of what Naomi Klein called "blockadia"—the worldwide grassroots movement to end the use of fossil fuels. Often the protagonists are frontline communities: Third World, Indigenous, poor, and subsistence peoples whose livelihoods are being destroyed by projects that extract or transport fuel for the benefit of the world's wealthy consumers. Divestment campaigns seek to undercut fossil fuel companies' legitimacy and financial viability by pressuring institutional and individual shareholders to divest. Both approaches, blockadia and divestment, aim to delegitimize, undermine, and stop the fossil fuel industry. Both work directly and concretely to prevent new fossil fuel extraction and use. Direct action and protest complement legal and financial strategies.

Many protests, like those targeting the Dakota Access Pipeline and terminals and compressor stations proposed in low-income areas already burdened with industrial and other sources of pollution, have been led by Indigenous and frontline communities. They emphasize the role of environmental justice in the struggle against fossil fuels.

Yet such protests have provoked controversy from different quarters. Some of them pit protesters against workers and unions that prioritize the jobs the projects will create. As I'll discuss in the next section, some activists argue that focusing on the fossil fuel industry exacerbates the labor-environmentalist divide by enabling industry to attack environmentalists as job killers. Instead, they advocate building labor-environmental unity by focusing on job-creating infrastructure investments.

Critics suggest halting one fossil fuel project may just lead to the growth of others. When Massachusetts succeeded in halting two pipeline projects in 2016 and 2017, the *Boston Globe* witheringly reported that the state was instead receiving LNG imports from a new Russian gas development project in the Arctic, with even more pernicious environmental consequences. The newspaper decried "the state's inward-looking environmental and climate policies" and "pipeline absolutism" with "scant consideration of the global impacts of their actions and a tacit expectation that some other country will build the infrastructure that we're too good for."[41]

In accusing pipeline activists of NIMBYism—saying "not in my backyard" to fossil fuel infrastructure and pushing it into someone else's backyard—the *Globe* posed a false either/or proposition: new pipeline infrastructure versus LNG imports. It took for granted that we must continue consuming more natural gas. The editorial evaded the very questions that those who care about the climate should be raising: how to achieve major reductions in energy use.

Most divestment campaigns pressure institutional divestment, especially by universities, pension funds, and religious institutions, from privately held fossil fuel companies. Some critics say divestment tips the balance in favor of the even larger, state-held companies like Saudi Arabia's Aramco, Russia's Gazprom, and Brazil's Petrobras, which control 90 percent of world oil reserves. Others say divestment's impact is limited because other investors will simply step in. Furthermore, fossil fuels are so deeply tied to every sector of our industrialized economies, it's hard to imagine where divested funds could be reinvested that would not continue to prop up the industry. Finally, even if divestment and other pressure campaigns succeed in convincing a particular oil company to clean up its operations, it's likely to do so, as British Petroleum did, by selling its dirtiest components to other companies, often privately held companies not susceptible to shareholder pressure.[42]

Clearly, divestment alone will not dismantle the fossil fuel industry. For most divestment activists, though, it's a part of a larger campaign. "Fossil fuel divestment takes the fossil fuel industry to task for its culpability in the climate crisis," explains 350.org's Go Fossil Free campaign. "By naming this industry's singularly destructive influence—and by highlighting the moral dimensions of climate change—we hope that the fossil fuel divestment movement can help break the hold that the fossil fuel industry has on our economy and our governments."[43]

It's true that approaching divestment or the halt of a pipeline as an end in itself allows us to fall into the trap of personal or institutional purification. But it doesn't make sense to attack divestment or pipeline protests on the grounds that neither one alone will stop fossil fuel production. No single strategy or policy will. As the IPCC emphasized in 2018, we need drastic interventions on multiple fronts.[44]

Divestment and pipeline protests have radicalized a generation of climate activists who have placed climate justice on center stage. If we can situate our individual actions and campaigns in terms of these larger goals, we will be better positioned to collaborate and make our campaigns complementary rather than insisting on the moral superiority of one approach over another.

DO WE NEED TO CONSUME LESS?

Yes, we need to consume less! People in the United States are among the biggest consumers of practically everything, which is what makes us the biggest consumers of fossil fuels.

If we want to figure out how to consume less as a society—not just as an individual—we need to understand the systems and structures that perpetuate our treadmill of consumption. Individuals in the high-consuming countries are simultaneously perpetrators *and* victims of the system. The solution is not just to reduce your individual consumption, it's to challenge the system that relies on our consumption and consent to keep it going.

Numerous studies have shown that our consumer mentality and society are not even necessarily in our own self-interest. Economist Juliet Schor suggested the concept of plenitude to replace the strictly consumption-based "standard of living" approach that dominates our society. Plenitude, for Schor, downplays endless accumulation in favor of more control over our time (working fewer hours), self-provisioning outside the market, an environmentally aware rather than purely market-based approach to consumption, and the rebuilding of community and human connections. As she summarizes: "Work and spend less, create and connect more."[45]

Working and spending less may sound more attractive to a high-powered, overconsuming lawyer than to someone struggling to make ends meet in a part-time, minimum-wage job. And even the lawyer may be worrying about putting kids through college and planning for a secure retirement. A corollary to our capacity for imagining plenitude in a lower-consumption lifestyle is a robust social safety net that takes basic needs and rights out of the insecurity of the market. If we knew we could rely on free daycare and college, real national healthcare and pension systems, and the security that our basic needs like food and housing were considered human rights and guaranteed by the public sector, it would be a lot easier to contemplate working fewer hours and redefining a quality of life that was not based on ever-increasing consumption.

I'll go into more depth as to what this kind of world could look like in the last two sections of this book. But for now, let's just note that consumption, profits, and fossil fuels are interrelated components of our current economic system. In contrast, many quality-of-life improvements (say, expansion of free education, public transit, and childcare) are low-carbon and have little potential for profit-making, even if a good portion of the population may want them. The private sector is good at developing what rich people want, because that's how it makes money. And the more excessive the consumption, the higher its emissions.

It has become a truism that the millennial generation in industrialized countries will be the first to fail to achieve a higher standard of living than their parents.[46] Most studies, though, define standard of living strictly in market-based and consumption terms. The millennial generation is also the first generation to experience the ravages of neoliberalism and the dismantling of the social safety net that complemented the consumer-based market economy. Millennials also have to deal with the climate and environmental consequences of their parents' profligacy. If we can acknowledge the environmental impossibility of ever-increasing consumption, and redefine quality of life in terms of public guarantees that allow us to get off the treadmill, consuming less can be transformed into an opportunity rather than a sacrifice.

CONCLUSION: WHAT KINDS OF INDIVIDUAL ACTIONS CAN MAKE THE MOST DIFFERENCE?

The kind of individual actions that can make the most difference are actually the ones that transcend the idea of acting as an individual. Even if you could adjust your individual behavior so as to radically reduce your own carbon footprint—stop eating meat, stop flying, stop driving, use solar panels to heat and/or cool your living space, give up your electronics, don't have children—you'll never get it down to zero. And even if you did, reducing the emissions of one of the planet's eight billion people, or even one of the planet's three to four billion high-consuming middle class, is little more than a drop in the bucket.

That doesn't mean that we should all just consume recklessly. Considering and understanding the role each of us plays as a cog in a larger system that exploits, produces, and relies on us (in the United States) to consume can be an important step in challenging the system. But emphasizing individual lifestyle change, without actually challenging the system, isn't going to get us very far. Self-improvement can be deeply depoliticizing if it encourages us to look inward rather than to confront society-wide problems. And climate change is quintessentially a society-wide, or planet-wide, problem.

CHAPTER 4

SOCIAL, RACIAL, AND ECONOMIC JUSTICE

Climate change is, of course, an issue for all people, especially poor people, who are more likely to live in the world's vulnerable areas and less likely to have the resources for private fortresses. Poor people also hold less power to enact policies that protect their interests. And, the global history of European colonialism means global poverty is racially inflected: the poor are more likely to be people of color, and people of color are more likely to be poor.

Scholarship and activism in environmental justice have shown the many ways that poor and working people in the United States and beyond have been disproportionately relegated to environmentally fragile regions, subjected to toxic industries and waste dumps, and forced to work in dangerous and unhealthy environments. They often lack access to clean food, water, and air. Their lack of economic resources and political power leave them more vulnerable to environmental hazards of all sorts.

Climate justice looks at climate-related expressions of racism and inequality and calls for changes in the structures that cause these problems. It locates the causes of climate change in an extractive economy that places profits before people and the planet. Climate justice recognizes that those least responsible for climate change, like Indigenous peoples and the global poor, are also those who are most vulnerable to its effects, and those who most lack the resources to protect themselves from extreme weather, rising seas, famine, and other environmental disasters. They are also those who have suffered the most from fossil fuel extraction, which continues to displace them as it poisons and destroys their lands. And they have benefited the least from the advances of industrial society enjoyed by the world's wealthy.

A climate justice approach asks how the social and economic divisions that characterize our world relate to the causes of climate change and its impacts. How have our institutions and structures enabled an elite minority to overextract, overproduce, and overconsume to the extent that their squandering of resources threatens human survival, while the consequences fall disproportionately on the global poor? The concept of climate justice suggests a need to look beyond technological fixes, policy tweaks, market mechanisms, and individual lifestyle adjustments to address the roots of the problem. Climate justice means recognizing climate change as a moral, political, and economic issue requiring a fundamental reorganization of our society and economy, not just manipulating incentives and enhancing technologies.

Because voluminous scientific research documents the human causes and disastrous impacts of climate change, and because right-wing climate deniers have relied on ignoring or dismissing this knowledge, progressive climate activists have rightly insisted that we "follow the science." But when it comes to policy, science can only take us so far. Science can analyze the physical, material aspects of the problem of CO_2 emissions, but it tells us less about the political, social, and economic aspects. This chapter argues that social justice issues, from race and racism to global economic inequality, are at the heart of the climate crisis.

WHAT'S THE RELATIONSHIP OF INEQUALITY TO CLIMATE CHANGE?

Back in the late 1980s, I attended a talk by one of the founders of the discipline of ecological economics, the Catalan economist Joan Martínez-Alier.[1] He began by saying he had just been contracted by the World Bank to conduct research on how Third World poverty contributed to environmental destruction. The idea was that poor people who lacked access to basic resources were more likely to cut down forests for cooking fuel, less likely to have access to sanitary latrines, waste disposal systems, and other infrastructure for environmental protection.

"But really," he told the audience, "I thought it would have made more sense for the Bank to do a study about how *wealth* contributes to environmental destruction." Not only do rich people consume more of virtually every resource than do poor people, but rich people control the global economy and the institutions that have overseen and profited from this global system based on ever-increasing consumption and destruction of resources.

Inequality characterizes the global economy, and the domestic economy of pretty much every nation. Rich countries—the former European colonial powers and their settler colonial offspring like the United States, Australia, New Zealand, and Canada, along with a few small outliers like the oil-producing countries of the Middle East—grew rich mostly by exploiting the land, labor, and resources of their colonies (which are today poor countries) and using those advantages, along with fossil fuels, to develop their own industry and increase their own consumption and emissions.

Every one of these wealthy countries maintains internal colonies of poor people, often defined as racially or ethnically different from the ruling majority and consigned to sacrificing their resources and their labor for those at the top. In the United States, these internal colonies include Native Americans, Blacks, and many immigrants from poor countries. In the high-consuming, oil-rich Persian Gulf states, migrant workers, mostly from South Asia, comprise more than half the population, and 95 percent of the workforce in low-wage jobs like construction and domestic service.[2]

Everywhere, wealth came from finding ways to intensify production, thus consumption, which meant using more resources and increasing pollution and emissions. Everywhere, wealth meant dispossessing others of their land, resources, and autonomy.

Of course, the technological advances of the past two hundred years have brought many positive changes to people's lives. People who have access to industrial society's goods don't want to give them up, while those without access desperately desire them. Yet it's also abundantly clear the planet cannot remain habitable if the global elite continue to consume resources at their current level, or if the global poor increases their consumption to anywhere near that of the global elite.

Overconsumption of resources by the few creates dangerous pressure on the other planetary boundaries, as identified by the Stockholm Resilience Centre. A more just and equal world would use its scarce resources at a sustainable level and distribute the benefits more equally. It would allow the global poor access to the resources necessary to fulfill basic needs and obligate the global elite to reduce their consumption—and emissions and resource use—to a just and sustainable level. Instead of increasing production and consumption overall, we'd need to work on reducing overall levels while redistributing the benefits.

While technology must play a role in this process, it is more a question of political and moral change. We must redefine basic needs as a collective good. We must confront the political and economic systems imbedded in our industrial revolutions and transform them to place the interests of humanity above the interests of profits and increasing production.

This does not mean that we should go back to preindustrial ways of living, before the medical, technological, and scientific advances of the past two hundred years. The problem is how they've been used: to concentrate wealth, consumption, and well-being among the planet's elite. Restructuring how we use our knowledge, science, and technology in a way that is equitable and sustainable is the political and moral problem that now faces us. We need to focus our human ingenuity and political systems on distributing resources more fairly and living within our planet's limits rather than on producing more.

WHAT DO RACE AND RACISM HAVE TO DO WITH CLIMATE CHANGE?

It might seem counterintuitive to say that racism, racial inequality, and climate change are so deeply intertwined as to be inextricable. And yet if we look at the history of how racial inequality came to characterize our country and our planet, it becomes clear that the processes of colonialism, extraction, and exploitation that created our racial order are the very same processes that created the climate crisis. These processes, led by white European countries over the past five hundred years, underlie global racial inequality and the consumption and living standard inequalities that accompany it. They also put colonized and formerly colonized populations, that is, people of color, who have contributed the least in terms of global emissions, on the front lines of the heat, drought, rising seas, and storms caused by climate change.

When we talk about racism, we often think first of individual behavior, hateful language and slurs, or legal policies of exclusion or segregation. Some have argued the United States is now in a post-racial era in which segregation has been outlawed and racist language is largely delegitimized, at least outside of far-right circles.

In fact, though, racism is woven so deeply into our history, society, and culture that it infuses virtually every aspect of our lives, even if we are taught not to see it. European colonizers elaborated ideologies of white European supremacy as they scoured the planet for resources. They justified abuse of other peoples' lands and labor by contending that "primitive" groups did not make productive use of their resources, that European ideas, institutions, and technologies were more advanced and could be used to increase production and bring progress to all (at least all who survived the holocaust of colonization). They justified enslavement, displacement, extraction, and colonial rule in the name of progress.

Racism in today's world is visible in the fact that Europeans, European colonizers, and their descendants tend to consume far more fossil fuels than do people of color. It's visible when Central American immigrants flee their violent, drought-parched, and exhausted lands only to be imprisoned at the US border, or when African Americans are incarcerated at five times the rate of whites, or when Native American and Black household net worth is less than a tenth of white households. When we are taught to overlook the causes of these disparities in the practices and history that made the country and the world we live in today, that's another manifestation of racism.[3]

The practices and history that created our racially divided world also created our climate emergency. Many of us in the United States have long clung to the comforting myth that, with a little more time, the benefits of industrial society could spread to those who are unfortunately still excluded. But the climate emergency should give the final lie to that myth.

We can't deny the privileges that industrial society has brought to its beneficiaries. But we can, and must, acknowledge its human and environmental costs. Confronting climate emergency means confronting racism and requires us to rethink the ideologies and beliefs that underlie our history—and our global economy.

HOW WILL DIFFERENT PEOPLE—AND DIFFERENT PARTS OF THE WORLD—BE AFFECTED BY CLIMATE CHANGE, NOW AND IN THE FUTURE?

Globally, poor people have been relegated to lands that are less fertile and more vulnerable to climate-caused disasters. They have poorer infrastructure and fewer resources to protect themselves and to recover. Living closer to the edge of survival, they are less able to withstand losses or rising prices. They have less political clout, so are less able to influence government policy.[4]

The warming climate contributes to extreme weather-related events like heat waves, tropical storms, wildfires, and floods, and to long-term changes like drought, rising sea levels, loss of water sources, desertification, pest and disease outbreaks, and shifts in agricultural capacity. All of these undermine livelihoods, especially of the rural poor in poor countries. Security analysts call climate change a "threat multiplier" because these stresses also contribute to armed conflict and mass migration. Rich countries provide the arms and barricade themselves against the migrants.[5]

The destruction caused by extreme weather events in poor regions opens the door to what Naomi Klein called "disaster capitalism," as politicians, aid agencies, and private capital use the destruction to rush in with their agenda of privatization and neoliberal reform. After Hurricane Mitch in Central America (1998), Hurricane Katrina in the US Gulf Coast (2005), or Hurricane Maria in the Caribbean (2017), impoverished governments invited the private sector to turn recovery into the chance to build an investor's paradise, decimating public services and profiting from the suffering. Disaster capitalism, then, means "orchestrated raids on the public sphere in the wake of catastrophic events, combined with the treatment of disasters as exciting market opportunities."[6] Carbon trading systems that create speculation and investment opportunities out of the longer-term disaster of climate change itself offer another form of disaster capitalism.

As we've seen, poverty and race are also linked. Africa, the world's poorest region, has been plagued by repeated droughts, tropical cyclones, flooding, and landslides. "Climate variability and change are among the key drivers of the recent increase in hunger on the continent," concluded an exhaustive report by the World Meteorological Association in 2020. Because most of its people rely on agriculture, Africa comprises "an exposure and vulnerability 'hot spot' for climate variability and change impacts" and "warming scenarios will have devastating effects on crop production and food security" there.[7]

Small island nations, especially in the Caribbean and the South Pacific, are particularly vulnerable to rising sea levels; some are threatened with complete inundation. Like Africa, almost all have suffered centuries of colonialism, extraction, and exploitation, have contributed little to fossil fuel emissions, and lack the resources to protect themselves against climate change's impacts.

In the poorest countries of South and Southeast Asia, the vulnerability that comes from heavy reliance on agriculture is compounded by the hundreds of millions of people living in inadequate conditions in overcrowded coastal cities. A significant majority of the world's one billion people likely to face lethal heat waves in the coming decades live in Asia.[8]

Latin America faces melting glaciers, an upsurge in tropical storm frequency and intensity, rising sea levels, drought, and species loss. Rising temperatures are spreading the reach of tropical diseases and, along with loss of water availability, decreasing crop yields, threatening especially those who work in agriculture (up to 40 percent of the population in the region's poorest countries). Because Latin America relies heavily on hydropower, loss of water sources will also affect the availability of electricity.[9]

Inside the United States too, poor people are now, and will in the future, be those who suffer the most from climate change. "Risks are often highest for those that are already vulnerable, including low-income communities, some communities of color, children, and the elderly," concluded a major US government study in 2018. Furthermore, "climate change threatens to exacerbate existing social and economic inequalities that result in higher exposure and sensitivity to extreme weather and climate-related events and other change."[10] The impacts of and chaotic and inadequate government responses to Hurricanes Katrina and Maria, as well as to the COVID-19 pandemic (the climate aspects of which are described below) offer vivid examples of the compounding effects of inequality and climate vulnerability.

Finally, our concern for social justice should extend to the rights of future generations. Today's children, and new generations as-yet unborn, will suffer the effects of our profligacy, perhaps forever. They have no political voice or power, but it's up to those of us currently alive to take their interests—and their very survival—into account.

HOW ARE PANDEMICS RELATED TO CLIMATE CHANGE?

Pandemics—like COVID-19—are related to climate change because the emergence of new pathogens stems from the same fundamental cause as our changing climate: the increasing pressure that industrial society is placing on the natural world. And both pandemics and climate change, despite being apparently neutral, impersonal phenomena that don't discriminate, intersect with existing inequalities and vulnerabilities to disproportionately harm the poor, people of color, and others without political and economic clout. The rich can insulate themselves from the full impact of these human-made threats or even profit from them; the poor cannot.

Finally, pandemics are related to climate change in that while science and technology are crucial to understanding them and changing their course, they are not enough. Science brought us COVID-19 vaccines, but wealth and power shaped who got access to them. We need to go beyond science to explain the political, social, and economic factors that keep us on a self-destructive path towards new pathogens and more climate chaos.

Most of the new diseases that have emerged among humans since the late twentieth century followed the same route: they migrated from animals to humans. Two building blocks have to be in place for this migration to occur. One, the pathogen must possess, or acquire through mutation, the capacity to infect humans. And two, humans must come into close contact with infected animals. Land use change, including deforestation and the expansion of human/industrial activity into previously forested areas, and industrial agriculture that concentrates huge numbers of potential hosts in close proximity, are two sides of the same coin, contributing to both of these prerequisites.

Even as vaccines, antibiotics, and public health measures succeeded in controlling many of the infectious diseases that had plagued humans for centuries, a plethora of new viruses and mutations started to appear with intensifying rapidity as the twentieth century drew to a close. They included outbreaks of HIV/AIDS in the 1980s, SARS1, H5N1 "bird" flu and H1N1 "swine" flu in the first decade of the twenty-first century, and MERS, Ebola, Zika, and COVID-19 in the second decade. Well before COVID-19 gripped the world, evolutionary biologist Rob Wallace had explained somberly that "humans have built physical and social environments, on land and in the sea, that have radically altered the pathways along which pathogens evolve and disperse." One of the most important human factors is agribusiness which, Wallace warned, "backed by state power home and abroad, is now working as much *with* influenza as against it."[11]

We've seen how enmeshed agribusiness is with the fossil fuel industry and what impact industrial agriculture has on land use and emissions. Factory farming aims at mass production, monoculture, and global production and distribution chains. As agribusiness expands globally into forested areas, it creates new environments for transmission from wildlife to domesticated animals and humans. Its "genetic monocultures" increase the chances that one of the small mutations that randomly occur in pathogens will reproduce exponentially if the mutation facilitates infection of a particular host. Concentrated animal feeding operations (CAFOs), which keep these genetic monocultures in constant physical proximity, help diseases spread, and the antibiotics they use to control the spread help promote the survival and reproduction of drug-resistant mutations. And the system's global commodity chains enable pathogens to spread quickly.

A key parallel between climate change and pandemics lies in the ways that we think about causality. We've already looked at proximate (the burning of fossil fuels) and structural (our economic system) causes of climate change. To truly understand the causes of pandemics we need to think beyond proximate causes like individual pathogens or, in the case of COVID-19, the details of its emergence in Wuhan, China. "The cause of COVID-19," Wallace argues, lies also "in the field of ecosystemic relations" that's been shaped by "capital and other structural causes."[12]

Fossil fuel companies don't actively desire climate chaos, and agribusiness doesn't actively desire pandemics. But they and their allies do actively promote business models and economic systems that cause CO_2 emissions and enable the emergence and spread of disease. "Attempts to proactively change poultry and livestock production in the interests of stopping pathogen outbreaks can be met with severe resistance by governments beholden to their corporate sponsors. In effect, influenza, by virtue of its association with agribusiness, has some of the most powerful representatives available defending its interests in the halls of government," writes Wallace. "In covering up or downplaying outbreaks in order to protect quarterly profits, these institutions contribute to the viruses' evolutionary fortunes. The very biology of influenza is enmeshed with the political economy of the business of food."[13] And like the fossil fuel industry, agribusiness is able to externalize these costs onto society as a whole.

Without negating the importance of vaccines and cures, Wallace insists that scientists' heroic attempts to deal with the fallout should not distract us from addressing the root of the problem: the agribusiness modus operandi. "Bird flu," he wrote in 2007, "now emerges by way of a globalized network of corporate poultry production and trade, wherever specific strains first evolve. We must devolve much of the production to smaller, locally owned farms. Genetic monocultures of domesticated birds must be diversified back into heirloom varieties, as immunological firebreaks. Migratory birds, which serve as a fount of influenza strains, must be weaned off agricultural land where they cross-infect poultry. To do so, wetlands worldwide, wildfowl's natural habitat, must be restored."[14]

A careful reader will note these are precisely the kinds of structural changes that can scale back our pressure on other planetary boundaries, including GHG emissions, and enable us to rebuild a global system that places the needs of the poor before the profits, speculation, and consumption of the rich.

Like climate change, the COVID pandemic exacerbated global and domestic inequalities and forced them onto the front pages in new ways. The job-related perils already present in sectors like agriculture, meatpacking, and elder care were exacerbated by the fact that these types of work place people in close quarters, tend to be nonunion, to lack benefits like decent health insurance, and generally leave workers with few resources for challenging unsafe conditions. Lack of personal protective equipment (PPE) and lack of interest on the part of employers and government made exposure to the virus an everyday hazard. Poverty wages mean crowded living conditions, facilitating the spread of infection at home and at work. People of color who are concentrated in these jobs are also more likely to have underlying health conditions, though their higher death rate is mostly attributable to a higher exposure rate.[15]

Such jobs were deemed "essential," yet their workers have consistently been treated as expendable. Large swaths of the US population noticed for the first time how essential these workers were and how perilous their jobs and living conditions were. "While essential workers are keeping others safe, they and those close to them are getting sick and dying," said SEIU (Service Employees International Union) 32BJ president, reporting that 132 members of his local had succumbed to the virus early in the summer 2020 surge.[16]

In poor countries, much larger proportions of the population lived and worked under these kinds of conditions, making lockdowns unrealistic or impossible and access to safety precautions and healthcare that much more difficult. Mexico's president, noting that 60 percent of the country's workers relied on daily earnings to survive, declined to mandate a lockdown. In India, Brazil, and other poor countries, health systems collapsed under the weight of raging infections.

Not only were the poor and people of color far more likely to contract and die from the disease, they were also far less likely to have access to the vaccines that were released in early 2021. A study in Boston found that 46 percent of white residents had a vaccine clinic within a mile of their home, while only 14 percent of Black residents did. Meanwhile high-income countries hoarded vaccine doses, immunizing large portions of their populations by mid-2021 while the virus raged in poorer parts of the world. "Rich countries enjoy the benefits and safety of being fully immunized, while people in poorer countries continue to get sick and die from the coronavirus," lamented one study. Most of Africa had little hope of obtaining vaccines before 2023.[17] A global map shows remarkable overlap between fossil fuel consumption and vaccine access among the world's wealthy and between vulnerability to the impacts of climate change and vulnerability to the pandemic among the world's poor.

HOW CAN WE FAIRLY HOLD DIFFERENT COUNTRIES, PEOPLE, AND INSTITUTIONS ACCOUNTABLE FOR THEIR CONTRIBUTIONS TO CLIMATE CHANGE? WHAT METHODS FOR CALCULATING EMISSIONS BEST SHOW WHO IS EMITTING THE MOST, AND WHERE TO TARGET OUR EFFORTS FOR CHANGE?

Numbers might seem to be the ultimate in objective facts. But decisions about what exactly we need to measure can hide, or expose, inequalities. Decisions on how to count emissions are political ones, with social justice implications.

Almost all official emissions statistics are calculated in terms of countries. International climate agreements, negotiated by representatives of the world's countries, have required emissions reporting and national plans to reduce emissions on a country-by-country basis. China is the world's largest emitter, the statistics tell us, responsible for 27 percent of global emissions in 2017. It's followed by the United States (15 percent), the European Union (9.8 percent), India (6.8 percent), Russia (4.7 percent), and Japan (3.3 percent).[18] If the European countries are counted separately, each individual country drops below the top six. These numbers indicate the biggest culprit is China, leading many to conclude that China bears the greatest responsibility for lowering its emissions.

Other ways of enumerating emissions lead to different conclusions about responsibility, like counting per capita emissions or cumulative emissions. Calculating emissions per capita (i.e., per person) highlights a country's consumption levels. China has the world's largest population, so as a whole the country is responsible for a lot of emissions. But most of its people individually consume and emit far less than do those in smaller but wealthier, higher-consuming countries.

Cumulative or historical emissions statistics measure how much a country has emitted over time, since it's the total amount of CO_2 in the atmosphere that causes climate change, not what's emitted in a particular year. Counting cumulative emissions shows that the United States and the European Union have by far the greatest historical responsibility, while emissions from China and India only recently began to rise as the countries have industrialized and their middle classes have grown. From this perspective, the United States and Europe owe a huge climate debt to the world.

Most statistics, like those required by the Paris Agreement and reported by the US Energy Department, use territorial or production-based accounting, which means they count emissions at the point of production. Some scholars argue because of the importance of global trade and the ways rich countries have outsourced their production to poor countries, it makes more sense to measure emissions by what is *consumed* in a country. Wealthy, high-consuming countries should be held responsible for the emissions of everything they consume, even if they've outsourced the dirtiest aspects of their economies.

Or perhaps we should not attribute emissions to countries at all but to individuals, who should be responsible for their emissions wherever they live. We tend to think in terms of countries, the world's main political units. But every country includes a wealthy elite that consumes a lot more than the rest of the population. Just as poor countries consume less fossil fuel, overall and per capita, than rich countries, poor people consume less fossil fuel than rich people within every country. Shouldn't the globe-trotting, jet-setting elites be targeted for consumption reductions, no matter what country they live in?

Other approaches focus on the fact that the biggest emitters are not individuals but industries. Maybe instead of attributing emissions to countries, we should look at emissions from the operations of specific corporations and those who control and profit from them. After all, it's the fossil fuel and related industries that make the decisions and investments that are expanding the extraction, production, and sale of coal, oil, and natural gas.

Understanding the differences among these ways of calculating emissions can help us think about how to hold countries, people, and institutions accountable for reducing emissions.

Per capita reporting

National-level statistics downplay the fact that the size of the world's countries varies considerably. The tiny European country of Luxembourg, for example, produced about 10 million metric tons of CO2, or about 17.5 metric tons per capita, in 2016.[19] That puts Luxembourg eighth in the world for per capita emissions. Only in eight countries—Qatar, Montenegro, Kuwait, Trinidad and Tobago, United Arab Emirates, Oman, Canada, and Brunei; all except Canada tiny nations that are major oil-producers—do people produce more in emissions per capita.

In total emissions, though, Luxembourg (10 million metric tons a year) is number ninety-nine among the world's countries, between Ethiopia and Zimbabwe. In Ethiopia, per capita emissions are only one-tenth of a metric ton; in Zimbabwe, they're just over one-seventh of a metric ton.[20] But both countries have far larger populations than Luxembourg. So though the three countries produce about the same emissions, Luxembourg produces a lot more per person than Ethiopia and Zimbabwe. That fact is hidden if we only look at national-level statistics.

So who bears more responsibility for contributing to climate change—the very wealthy population of Luxembourg, with extraordinarily high per capita emissions or the very poor majorities in Zimbabwe and Ethiopia, many of whom consume little in the way of fossil fuels at all? If we want a fairer world, we should pursue policies that allow Zimbabwe and Ethiopia to consume more energy while people in Luxembourg consume less. If we want a habitable world, this shift has to be accompanied by an overall significant reduction in fossil fuel use. But to be fair or just, the reduction should focus on high-consuming Luxembourg, not on low-consuming Zimbabwe and Ethiopia, even if, according to national-level statistics, their emissions rates look similar.

The example of Luxembourg shows us one kind of distortion: when a very rich country with a very small population has what looks like a middling emissions level but is really producing far more than its fair share. Another way of exploring the importance of including a per capita measure is to examine poor countries with extremely large populations, like India and especially China.

If we look just at national-level emissions, China is by far the most culpable country in the world, with 10.5 billion metric tons a year. That's more than twice the total for the next-largest emitter, the United States. But each person in China emits, on average, only half of what each person in the US emits. China's per capita emissions stand at a middling 7.4 metric tons a year (forty-second worldwide), while US per capita emissions are 15.5 metric tons a year (sixteenth in the world). The first fifteen are much smaller countries than the United States, so their total emissions are much smaller.[21] Per capita reporting helps us see people's overconsumption much more clearly than does national-level reporting.

Historical responsibility and cumulative emissions

Instead of looking at how much CO2 a country currently emits each year, a historical perspective examines cumulative emissions, or how much CO_2 a country has emitted over time. Cumulative emissions are important because greenhouse gases in the atmosphere accumulate over time; they don't disappear at the end of the year they are counted in. That's why even if we stopped *all* emissions right now, the planet would continue to warm. The gases we've already emitted will stay in the atmosphere for decades or even centuries.[22]

When we use a historical perspective, the inequality of global responsibility appears even starker than annual total or even per capita emissions. The European countries and their industrialized offspring—the United States, Canada, Australia, and New Zealand, populated primarily by Euro-descended people—were built on the wealth of extractive colonialism, which they used to industrialize and build the fossil economy. They began emitting more than their share of CO_2 back in the 1800s and accelerated over the course of the 1900s as their industrial economies boomed.

More than half of the world's total emissions before 1882 came from the early and heavily industrialized UK. Until 1950, more than half came from Europe. US emissions also grew quickly as the country industrialized. The United States accounts for twice the amount of historical emissions as China, even though China emits more each year. In fact, the US now accounts for a quarter of all historical CO_2 emissions, while the twenty-eight EU countries account for 22 percent. Meanwhile the UK, despite its historical emissions debt, only accounts for about 1 percent of current annual emissions.[23]

Looked at another way, the big historical emitters are also the big per capita emitters. That is, they industrialized and developed by emitting a lot of CO_2, and per capita and consumption-based accounting reveal how the people of those countries are still reaping the "benefits" of this history by enjoying high consumption levels. Jason Hickel calls it "atmospheric colonization." The Global North not only appropriated the Global South's land, labor, and resources but also used up its share of the "global atmospheric commons." The entire world is forced to pay the climate price for the colonial industrialization of a few countries.[24]

Consumption-based accounting

Most national-level reporting uses a territorial or production-based accounting system, tallying the emissions produced by activities within the country. But a lot of what's consumed in wealthy, high-consuming countries is actually produced in poor countries. Should Bangladesh, or Vietnam, be considered responsible for the emissions of foreign-owned factories in their countries that are making products for export? Shifting our attention from production to consumption offers yet another way of apportioning responsibility for emissions.

A number of studies have used the concept of emissions embodied (or embedded) in trade to calculate how countries import and export emissions. Almost a quarter of global emissions come from the production of goods that are consumed outside of the country where they are produced. So these emissions comprise a significant contribution to the global picture.[25]

China, the United States, and Europe offer examples of how this works. Since much of what's consumed in the United States is produced outside its borders, it is a net importer of emissions. For China, it's the reverse: its factories produce emissions for items that are exported and consumed elsewhere. So the United States imports emissions embedded in goods, while China exports them. These embedded emissions would add 7.7 percent to US emissions as currently calculated. In China, 14 percent of its emissions come from production for export. Europe's calculus is even more extreme: imports would add some 30 percent to domestic emissions in many countries, and add up to three times domestic emissions in Switzerland.[26] The current international reporting system that shows only production-based accounting exaggerates the contributions of poor countries. Consumption-based accounting makes the responsibility of wealthy countries that import goods from poor countries more clear.

The role of the global elite

Some studies of consumption-based emissions disaggregate consumption within countries, and reveal what per capita statistics hide: in every country, some people consume a lot more than others. This approach offers another way to think about social justice and climate change. The highest emitters are concentrated among the ultra-wealthy classes that run business and industry and possess the means to influence policy, so looking at them also helps us analyze the structures they control and the ideas they promote.

Within every country, rich people contribute much more to emissions than poor people. Rich people buy more, consume more, drive more, and fly more. They own and invest in more industries, including extractive industries like fossil fuels. They live in larger homes, fill their homes with appliances, and are more likely to control the temperature of their homes and other spaces they move in. Whether we measure emissions by production or consumption, rich people in every country contribute far more than their share, and poor people far less. Consuming more, and producing more emissions, is built into the nature of wealth in today's world.

About 45 percent of the planet's emissions come from only 10 percent of its population—the high-consuming global elite. Meanwhile the global poor—about 50 percent of the planet's population—only contribute 13 percent of annual emissions. Almost half of the high-emitters live in the United States. One-third of the world's high emitters actually live in poor countries; they belong to those countries' elite classes.[27]

The top 1 percent of emitters—the richest of the rich, most of them located in the United States, Canada, Luxembourg, Saudi Arabia, and Singapore—emit over 200 metric tons of CO_2 equivalence per person every year. The lowest emitters—the poor of Honduras, Mozambique, Malawi, and Rwanda—emit two thousand times less, or about 0.1 metric ton a year.[28] Race plays a role here too. The global elite are disproportionately Euro-descended; the global poor are people of color.

China's apparently high national-level emissions numbers, and mid-level per capita numbers, also look very different when we take into account the country's vast inequalities. China's small wealthy elite and growing urban middle class are responsible for far greater per capita emissions than is its large population of rural poor. (Outsiders tend to conflate all inhabitants of China as "Chinese." But China, too, is multiethnic, and rural poverty rates among ethnic minorities are double those of ethnic-majority Han Chinese.[29])

One study showed that "income redistribution in urban China could reduce aggregate carbon footprints while improving living standards and income inequality. . . . Social and redistributive policies need to be understood as interacting with climate and energy policy, as well as with efforts towards enabling sustainable lifestyles for all." Few studies have attempted such analyses elsewhere. But even within the United States, the carbon footprint of the wealthiest Americans, the 20 percent that earn over $150,000 a year, is five times that of the poorest, who earn under $10,000 a year. In the United States, large suburban houses emit fifteen times more greenhouse gases than modest homes in low-income neighborhoods.[30]

If we define emissions in terms of individual consumption, we visibilize inequalities, which is important. But we shouldn't let the focus on individuals, even rich individuals, prevent us from understanding the institutional forces that enable these individuals to get so rich, consume so much, and keep the system working in favor of their ongoing consumption and emissions.

Financing emissions

Rich people—and countries—outsource their emissions when they import from, or invest in, factories that produce things elsewhere. They also outsource emissions when they finance and make profits from fossil fuel development in other countries. The biggest financers of coal and other fossil-fuel-based projects in poor countries—China, the United States, the EU, and Japan—are also some of the biggest emitters themselves. Private companies, banks, and investors work hand in hand with public finance agencies like the US Export-Import Bank, the US International Development Finance Corporation, and the European Investment Bank. Even countries that have banned new coal plants at home continue to finance them abroad. But the emissions from the new coal and other fossil fuel plants that they profit from in India, Southeast Asia, and Latin America get counted towards those poorer countries. Between 2015 and 2021, global banks poured $3.6 trillion into new fossil fuel projects, taking home billions in profits. Not until mid-2021 did US and European governments begin to discuss ending their support for such projects. Finance-based emissions counting would hold those financing and profiting from installations that create emissions responsible for them.[31]

The "Carbon Majors"

Yet another way of apportioning responsibility for emissions is to look at the companies that are extracting and selling fossil fuels. The top one hundred fossil fuel companies are responsible for over 70 percent of GHG emissions between 1988 and 2017.[32] Twenty fossil fuel companies together were responsible for a third of historical GHG emissions, as of 2019. Just four of them—Chevron, Exxon, BP, and Shell—together account for 10 percent. Twelve state-owned companies, led by Saudi Aramco, are responsible for another 20 percent.[33]

Whether state-owned or private, these companies rely on government largesse. They are quite adept at defending their interests and keeping us on the fossil treadmill. All of them benefit from government infrastructure and support.

The Climate Accountability Institute's Richard Heede explained starkly, "These companies and their products are substantially responsible for the climate emergency, have collectively delayed national and global action for decades, and can no longer hide behind the smokescreen that consumers are the responsible parties."[34]

How we count emissions matters

Different methods of calculating emissions encourage us to think in different ways about what kinds of changes we need to make to reduce emissions. The various international climate accords have thus far used national-level totals and focus on how each country can reduce its emissions using current emissions as a starting point. And they use production-based accounting. This approach makes a certain degree of sense, but it also naturalizes, by invisibilizing, some of the powerful actors setting the rules of the global economy and the terms of the debate. It naturalizes the drastic inequalities that characterize our world and assumes that these inequalities should and will continue.

Each method of calculating GHG emissions emphasizes different causes of these emissions. While they all may be equally valid in terms of the science and math behind them, they have very different implications in terms of policies and actions for change. Where we focus our attention is important, because it shapes how we think about paths forward.

How can we use the information offered by these different approaches to think about fair and effective ways of achieving the emissions reductions that are so urgent now? While governments are the entities best positioned to implement reductions policies, I'd argue that we should pay more attention to a country's per capita than its total emissions. The over a billion people in China shouldn't be required to use less per person just because they live in a large country, and the six hundred thousand inhabitants of Luxembourg shouldn't be able to continue their ultra-high consumption just because their country is small. In a way, tallying emissions by country rather than per capita for the purposes of international policy-making is just another colonial move that whitewashes the (generally smaller) European colonial powers.

In 2013, the IPCC proposed a "carbon budget"—the total amount we can emit globally to stay within the Paris Agreement's warming target of 1.5 or 2 degrees Celsius. Divided among the world's almost eight billion (or perhaps nine billion by 2050) inhabitants, to stay within the higher limit, that comes to about 2 metric tons of total emissions a year, or 1.6 metric tons of CO_2. Scientists warned that without a radical shift in direction, the planet's carbon budget would be used up within thirty years. Instead, emissions continued their inexorable rise. By 2018, some analysts believed that the remaining budget limited us to well under a metric ton per person per year. Each year that we fail to take action makes the task that much harder.[35]

We must also take into account the problem of cumulative emissions. The world's wealthiest countries, in particular the United States, would owe a huge carbon debt to the world for the damage they have caused with their emissions over the past two hundred years, even if they were to stop emitting entirely tomorrow. We've *already* used up far more than our fair share of the planet's carbon budget. A system of climate reparations could require historical emitters to compensate poor countries for damages up to today and facilitate their rise to a common, globally sustainable level of consumption by underwriting low-carbon development there. This is the kind of global climate justice that many grassroots organizations and the US Climate Fair Share Coalition are demanding.[36]

Countries are the fundamental policy-making entities in our world. But a social justice approach takes into account differences within and among countries and the role of governments, global elites, our global economic system, industry, banking and finance, corporations (in general), and fossil fuel companies (in particular) in creating and perpetuating the climate crisis. A fundamental weakness in the policies the world has attempted to implement thus far has been countries' reliance on these very groups to take the lead in addressing the crisis. But they're the ones who benefit from the system. As long as they are leading, they're unlikely to make real change.

WHAT DO WORKERS AND THE LABOR MOVEMENT HAVE TO SAY ABOUT CLIMATE CHANGE? IS CLIMATE CHANGE A UNION ISSUE?

It's not surprising that global elites and powerholders are reluctant to upset the foundations of a system that benefits them. But why have workers and their unions been so frequently estranged from the climate justice movement? A look at the US labor movement's history can help explain.

Unions are based on the idea that workers share a collective set of interests. Their work creates wealth, but in a capitalist economic system, much of that wealth flows upward into the pockets of owners and investors. By organizing together, workers can press for a fair share and a decent life for their members and for workers in general. The labor movement—individual unions and the federations they have formed—has been a key area for worker organization and worker power.

Unions have fought for rights like fair wages, benefits like healthcare and pensions, and safe working conditions. These fights take place at the level of a single plant or industry, where workers who have elected to be represented by a union negotiate a collective bargaining agreement with their employer or use the power of a strike to press for their demands, or they take place at the state or national level, where unions and their federations have pushed for legislation like unemployment insurance, child labor laws, minimum wages, workers' compensation for those who are injured on the job, and health and safety guarantees.

In the nineteenth and early twentieth century, many unions opposed capitalism. Anarchists, communists, and socialists believed the factory system was based on owners exploiting workers to make a profit. Owners would always want to pay as little as possible in order to keep as much as possible, and workers had to organize collectively to demand a larger share—and to change the whole economic system. Why shouldn't workers, or the government, run industries in the public interest, instead of private investors making the decisions and taking the profits?

Over the course of the twentieth century, a series of Red Scares succeeded in purging most unions of their more radical leaders and members. New Deal legislation established a framework of legal rights for workers, and the National Labor Relations Act set up a system for workers to hold elections to choose union representation and for unions to negotiate with employers for a collective bargaining agreement or contract. The union movement, joined together in the AFL (American Federation of Labor), and then after 1955, as the AFL-CIO (Congress of Industrial Organizations), came to accept capitalism.

The postwar economic order was based on unionization of key industries like steel, mining, and automobiles, with high wages and benefits for unionized workers. Unions bought into the notion of an American way of life based on home and car ownership and consumerism that granted certain sectors of the US working class access to the country's unprecedented fossil-based material prosperity. Unions' main work came to be negotiating and administering contracts, although they also maintained close ties with the Democratic Party and promoted pro-worker and pro-union legislation.

Union workers became a relatively privileged sector of the labor force, often with health, pension, vacation, and other benefits that nonunionized workers could only dream of. Critics accused unions of operating a "private welfare state" for their members and failing to work for the interests of the masses of unorganized workers. Supporters emphasized that by raising standards for organized workers and pushing for legislation, unions have benefited all workers.[37]

The postwar working-class consciousness and labor movement were racially and nationally inflected. The "American" way of life implied that non-Americans did not deserve the same access. Even within the United States, many low-wage workers, in particular people of color like farmworkers and domestic workers, were excluded. Racial and national parochialism formed the underside of this American dream.[38]

And so did fossil fuels and environmental destruction. The labor movement celebrated economic growth and ever-increasing production and consumption. Increasing production created more jobs; increasing consumption raised the standard of living. Each sustained the other, and they cemented unions' adherence to fossil fuels, the capitalist system, and economic growth. The mindset didn't leave much room for questioning the environmental sustainability of endless expansion on a planet with finite resources or what this pursuit meant for the people of the Global South.

This history offers part of the explanation for organized labor's fraught relationship with organized environmentalism. But the environmental movement's elite roots also played a role. Workers and their unions did care about the environment, and they've had particular class-based reasons for doing so. While mainstream environmental organizations sought to preserve wilderness from human interference, working-class environmentalism has focused on cleaning up the environments people live and work in and on enabling urban workers, many of whom were migrants from rural areas to industrial towns and cities, to have access to the natural world. Poor and working-class people are more likely to live in neighborhoods affected by environmentally hazardous industries and waste dumps and to lack clean air and water. From Progressive-era occupational safety and public health campaigns, to mid-century United Auto Workers–sponsored rural summer camps, to United Farm Workers campaigns to ban toxic pesticides in the fields, workers and their unions have been at the forefront of many environmental struggles in the United States. And globally, peasant struggles for land and food sovereignty against agribusiness and extractivism constitute another aspect of what Joan Martínez-Alier identified as an "environmentalism of the poor."[39] Yet the major US environmental organizations often seemed to prioritize rare animals and remote wilderness over the everyday needs of people and their jobs.

The late twentieth century: Climate crisis and assaults of labor

By the 1970s, the New Deal–era pact began to disintegrate as corporations began a new assault on organized labor. They hired union-busting law firms, pushed for neoliberal reforms that gutted the public sector and deregulated the private sector, subcontracted and relocated to escape unions, and threatened and pursued plant closures that shifted production abroad. Unionization rates plummeted as industries closed. The new service, high-tech, and gig economies that emerged at the end of the century were almost completely nonunion.

At the height of unionization in the middle of the twentieth century, unions represented some 35 percent of the US labor force. By 2020, this had shrunk to just over 10 percent. As unions have declined, so has workers' political power and their share of the national income.[40] Labor unions still comprise the largest organized voice of working people, even though the great majority of workers are not represented by a union.

As industries closed, relocated, and automated, workers were left jobless and mining and factory towns were devastated. As the climate crisis moved into center stage at the end of the last century, many unions were in a fight for their lives. Some unions responded by reviving the movement's left wing and social justice unionism. Others retrenched into nativism and protectionism, often allying with employers in fights against climate, environmental, and other regulation in what they saw as a last-ditch effort to save jobs.[41]

The AFL-CIO opposed the 1997 Kyoto Protocol, supporting the US position that it was unfair to require greater emissions reductions from the biggest polluters. Since then, the federation has consistently supported the US government's obstructionist position that wealthy countries should not be required to meet emissions reductions standards unless poor countries do too.[42] It's been extremely difficult for the AFL-CIO to relinquish its idea of an "American standard of living," which is implicitly a pledge to global inequality, and its commitment to the fossil fuel industry and economic growth. And it's difficult for unions (in general) and energy unions (in particular) to believe the kind of economic transformation that many in the climate justice movement are calling for will take workers' interests into account.

Fossil fuels and global justice

Particular divisions emerged around the issue of fossil fuels, carbon capture and storage (CCS), and global justice. Industry defends ongoing fossil fuel use and advocates public investment in CCS to make it cleaner, while banking on only cosmetic shifts in the global economic order. The climate left sees ending fossil fuels as an urgent priority and as part of a global economic transformation. The labor movement reflects the same divide, with the AFL-CIO leadership and energy- and building-sector unions tending to support ongoing fossil fuel production, new pipelines, and CCS, while some public- and service-sector unions ally with Indigenous and climate justice movements fighting against fossil fuels. The positions, while sometimes blurred and shifting, mirror the split in environmental movements between big green and climate justice organizations. In the policy arena, they map onto the differences between Biden's American Jobs Plan and radical versions of the Green New Deal.

Many environmental activists blame the labor movement for falling into traps set by the fossil fuel industry, what labor scholar and activist Sean Sweeney calls it the "black-blue alliance" (i.e., coal industry and blue-collar workers). But some labor activists fault the climate left's commitment to eliminating fossil fuels, and mobilizations against fossil fuel infrastructure projects, as unnecessarily contentious. Vincent Alvarez, president of the AFL-CIO New York City Central Labor Council, argued that "focusing on the 10 percent of the issues that are divisive—such as the Keystone pipeline and fracking" is a losing strategy for environmentalists. The fossil fuel industry uses these wedge issues to undermine both labor-environmental solidarity and the environmental movement as a whole. Instead, Alvarez suggests, "It makes more sense to start with the 90 percent of the issues that environmentalists and unions can easily agree on, including infrastructure, public transportation, energy production."[43]

The BlueGreen Alliance, which began as a collaboration between the United Steelworkers and the Sierra Club and now includes other major environmental organizations and unions, embodies this vision, uniting around good jobs, clean infrastructure, and fair trade. Its goals align closely with Biden's American Jobs Plan, including, controversially, support for carbon capture, implied inclusion of natural gas as a "clean" energy source, and a concept of "fair trade" that comes closer to protectionism than the restructuring advocated by Third World and global justice organizations.[44]

Many climate justice organizations argue that an infrastructure and trade plan that fails keep fossil fuels in the ground or address the global economic system will do little to avert the encroaching climate disaster. Radical voices in the union movement advocate concepts like global solidarity, energy democracy, and a just transition, urging unions to break their alliances with industry and challenge the ways that neoliberal fossil capitalism ravages both workers and the planet. Rather than supporting corporate-dominated economic arrangements while advocating for slightly greener capitalism, climate justice unionism could embed the struggle for workers' rights in a larger movement for public services and against privatization in a challenge to capitalism that offers what Sweeney terms "a new ecological and economic development paradigm." Public- and service-sector unions have generally been more open to this perspective than those in the energy and manufacturing sectors.[45]

To the left of the BlueGreen Alliance, the Labor Network for Sustainability (LNS), founded in 2009, offered an early (2016) and more radical framework for a Green New Deal when it advocated federal action "comparable to the economic mobilization that the US undertook during World War II" to wind down emissions in all sectors and commit to 100 percent renewable energy. In an explicitly climate justice framing, the LNS supports "a democratically controlled, sustainable, demilitarized, equitable, and just economy that uses our resources not for greed or military domination but to meet the needs of people and planet."[46] The call for government action and renewable energy overlaps with more centrist approaches, but the commitment to the end of fossil fuel use and to global solidarity places the Labor Network with the climate justice left rather than the big green mainstream. Fossil fuels and global justice are the fault line.

Reactions to the 2019 Green New Deal (GND) resolution highlighted the differences. The climate left and the LNS strongly supported it (the LNS also helped to frame the much more detailed and radical Green New Deal for Europe). The BlueGreen Alliance was silent, though some of its member organizations came out in support. The country's largest energy- and building-sector unions responded defensively on AFL-CIO letterhead, pleading that "the voices of American workers be included in the discussion, especially those who are most at risk of job disruptions and economic dislocations as a result of those actions." The GND proposal, they wrote, "is far too short on specific solutions that speak to the jobs of our members." The unions particularly lamented the proposal's lack of "an engineering approach," an implicit reference to CCS.[47]

This reaction led some environmental organizations, especially on the left, to dismiss the AFL-CIO as part of the problem rather than a potential ally. Friends of the Earth caricatured the AFL-CIO stance, unhelpfully likening the federation to "climate deniers like the Koch brothers, the Republican Party, and Big Oil."[48]

There is no single answer to how to build stronger relationships between labor and environmental movements, each of which contains a wide range of positions internally. A legacy of mistrust, the urgency of the climate crisis, unions' history of parochialism, and their current weakness in the face of legal and corporate attack that have seriously undermined the stability and futures of much of the US working class make it a difficult proposition. But the growing popularity of unions and their renewed commitment to social justice unionism, along with the urgency of climate action and the climate movement's increasing commitment to the needs of workers and the poor, offer hope for a path forward.

The Green New Deal Network and its February 2021 THRIVE Agenda, introduced by Senator Ed Markey and Debbie Dingell, US representative from Michigan, offered an example of how the different positions could come together under a Biden administration. The Green New Deal clearly influenced Biden's plan, and Biden's strong relationships with organized labor has helped bring more unions in line with the need to link union jobs, climate justice, and urgent climate policy. THRIVE goes beyond Biden's American Jobs Plan in its ambition ($1 trillion of investment a year), its commitment to end fossil fuel use, and its work bringing on board labor and environmental groups from the center and the left by consistently linking labor, climate, and racial justice goals.

WHAT IS A "JUST TRANSITION"?

A "just transition" is a key but controversial concept in the climate movement that highlights some of the disputes among the environmental, climate, labor, and other social justice movements.

The Labor Network for Sustainability locates the origins of the concept of a just transition at the end of World War II, when the government took steps to protect veterans in the transition from a wartime to a peacetime economy, providing education, home loan, and unemployment benefits to cushion the return of sixteen million servicemen and women to civilian life.[49] The rationale of the 1944 GI Bill was that it would be more just, and better for the economy, if the government took responsibility for the welfare of those adversely affected by the end of the war. Transition to peace was in everyone's interest, but structurally, those who had served in the war would be left vulnerable if the peacetime economy did not offer them a path to economic security.

The GI Bill was built on the sense of national purpose and the idea that prevailed during the New Deal and the war that government was responsible for the well-being of the population. It acknowledged that people who served in the armed forces had contributed more than their share, and at the same time were likely to suffer disproportionately from the transition. It was the government's responsibility to ensure that the transition was just for them.

One thing missing from this postwar consensus was attention to racial injustice. Cold War anti-communism led to attacks against radical, multiracial organizing inside and outside of unions. While the GI Bill itself did not impose racial discrimination, it didn't acknowledge the structural racism that shaped its impact either. If Black people and other people of color were excluded by segregation from neighborhoods, jobs, and higher education institutions, the education and home loan provisions of the GI Bill could serve to exacerbate, rather than overcome, entrenched racial inequalities.

Military spending was a form of job creation that could be counted on to meet with Congressional and public approval. Some called it "military Keynesianism" after John Maynard Keynes, an economist who argued that government social welfare would stimulate demand and prevent another depression. Critics of the military-industrial complex as the engine of the economy urged conversion to a peacetime economy with the slogan "Jobs with Peace." But they were drowned out by the Cold War. When the Cold War ended with the disintegration of the Soviet Union after 1989, there was a renewed push for a "peace dividend" that would transfer government spending to socially useful areas. But US military spending didn't really decrease, and it shot up even more after September 11, 2001. Meanwhile, neoliberal economic ideas calling for lower taxes and government cutbacks in social services prevailed over Keynesian economics and support for the social welfare state.

Among those peace activists pushing to "ban the [nuclear] bomb" and cut military spending in the 1970s was Tony Mazzocchi. Head of the Oil, Chemical, and Atomic Workers Union (OCAW, now part of the United Steelworkers), he revived the logic of the GI Bill, arguing that workers should not be forced to shoulder the costs if their polluting or harmful industries were shut down or transformed. Workers threatened by disarmament, Mazzocchi reasoned, deserved the same kind of support that demobilized soldiers deserved. In the 1990s, he pressed the idea that energy workers, who knew the true environmental costs of their industries better than anyone, could also lead the fight against climate change and for a just transition away from fossil fuels. He called for a Superfund for Workers, modeled on the 1980 Superfund program that designated federal funds to clean up areas contaminated with toxic substances.

"Paying people to make the transition from one kind of economy—from one kind of job—to another is not welfare," Mazzocchi explained, adding that workers who had built and sustained the energy infrastructure "deserve a helping hand to make a new start in life."[50] By the mid-1990s, the term "Superfund for Workers" evolved into "Just Transition," and expanded its scope as the OCAW began to reach out to frontline communities devastated or displaced by energy production, climate change, or other ravages of the underside of industrial society. Neither workers in toxic industries nor communities affected by those industries should pay the price for the transition, activists argued. The government must compensate those who had sacrificed in the past and would suffer again from the closure of an industry.

By 2000, the concept, adopted by the International Trade Union Confederation (ITUC), many climate and environmental justice organizations, and mainstream environmental organizations, was even incorporated into international climate agreements. And the concept plays an important role in the idea of a Green New Deal. The just transition has been a key tool for trying to overcome the rifts between environmental, community, and labor interests.

Despite its labor origins, much of the US labor movement remains deeply suspicious of the just transition concept. Many, like former AFL-CIO president Richard Trumka, believe it's no more than a euphemism for "a fancy funeral," a token that will never compensate for industry closure. Workers have been the victims of economic transition over and over again since the 1970s. As US coal mines and industries have shuttered, the transition has never been anywhere near "just." Instead, it has led to jobless workers and devastated communities. When asked about the concept, the United Mine Workers of America president replied bluntly, "I've never seen one." Brad Markell of the AFL-CIO noted that one quarter of the coal industry had already closed down, and "there's been nothing for coal workers. . . . Workers have been getting the short end of the stick of decades, and this of course colors how they see any big changes in our economy."[51]

It's still a struggle to find a common vision of a just transition that fulfills the dreams of both workers and affected communities. Environmental justice communities—whether it's Native Americans in North Dakota fighting oil pipelines, Black families living among Louisiana's "Cancer Alley" petrochemical industries, or Kentuckians contesting mountaintop removal coal mining—tend to oppose the very corporations that workers look to for jobs and pensions, even if many workers also live in affected communities.

Biden's jobs plan, like the Green New Deal resolution, emphasizes the need to prioritize both workers and "frontline communities" affected by climate change. Many in the climate and labor movements are working on articulating visions of what this just transition could look like. So far, the idea remains mostly aspirational.

WHAT IS "ENERGY DEMOCRACY"? HOW IS IT RELATED TO THE STRUGGLE TO CONFRONT CLIMATE CHANGE?

The call for energy democracy is based on the belief that energy is a basic human need and human right. Our current market-based energy system is geared towards producing and selling as much fuel as possible to those who can best afford it, without regard for social and environmental costs. A very small number of powerful individuals and institutions set the rules and make the decisions, while the vast majority of the world population has no say at all. Control and exploitation of concentrated energy resources in the form of fossil fuels is at the heart of race and class divisions today; this control created wealth and poverty, development and underdevelopment, and of course, the growing climate disaster. The costs are currently paid by the poor, whether in the oil fields of Nigeria or in the drought-ravaged countries of Central America, while the powerful enjoy the benefits. This status quo is the antithesis of democracy.

Energy democracy means wresting decision-making power from those who control the system: profit-seeking corporations, governments, and international institutions that appropriate energy resources for the benefit of the few. It means claiming energy as a common good and a human right, and developing ways to assert popular control over the decisions that affect us all. "[The] energy commons," writes climate scholar and activist Ashley Dawson, "must be about more than simply switching from fossil fuels to solar power: at its heart, this struggle must enable radical redistributions of power that don't just democratize but also effectively decolonize energy and society."[52] True democracy goes beyond elections and beyond national borders: it means transforming a global economic system to limit energy overuse by the wealthy while prioritizing the energy needed to fulfill basic human needs.

One way to do this is to take energy resources and decisions out of private hands and put them under public control, although government control alone doesn't ensure lower emissions or fairer energy distribution. State-run oil and gas companies like Saudi Aramco or Russia's Gazprom are not that different from publicly traded corporations like BP or Exxon in their drive to make money by exploiting, producing, and selling fossil fuels with little regard for the social and environmental costs. Some oil-producing countries offer hefty subsidies on gasoline and utilities, turning their populations into heavy consumers and amounting to another gift to the industry. State-owned fossil fuel producers, like private companies, seek to externalize costs, maximize profits, and evade regulation and taxes.

Even in a highly democratic country like Norway, a state-run oil industry can create perverse incentives to maximize fossil fuel extraction. Norway is a global leader in employing domestic policy to keep fossil fuel use low, at least in electricity production, while it continues to extract and export petroleum.

"Norway is almost entirely run on hydropower, which meets about 95 percent of the country's energy needs," writes the author of a 2018 study. "New buildings larger than 500 square meters (about 5,000 square feet) are required to get 60 percent of their energy from a renewable source. Cities have vast green spaces, bike lanes, and little traffic. Oslo's city center is [close to] car-free. And if one must own a car, it should be an electric vehicle. Norwegians who own them get breaks on parking fees, tolls, and more. All this means that, by some measures, Norway is among the world's greenest countries." Yale University's Environmental Performance Index ranks it fourteenth in the world for its sustainable practices.

Still, Norway's emissions, mostly from transportation and industry, come to 8.3 metric tons per year per capita. That's significantly lower than US per capita emissions, but it's well above its neighbors the UK, Italy, and France and far exceeds the 1.6 metric ton maximum necessary to reach the goal of remaining within 1.5-degrees Celsius warming. (Some studies using consumption-based accounting place Norway's per capita emissions much higher, at over 17 metric tons.)[53] Furthermore,

Norway is the world's fifth-largest per capita petroleum extractor; oil accounts for half of the country's exports. Oil profits fund the country's extensive social welfare system, and the Government Pension Fund Global of Norway, the largest sovereign wealth fund in the world. If we count the emissions from the petroleum Norway extracts and exports, the country falls to 128th place in Yale's index.[54]

So democratic Norway does a great job of meeting its citizens' needs and in some important aspects protecting the environment at home, but is more problematic in terms of its contribution to global emissions. (On the other hand, Norway produces only about 2 percent of the world's oil, and its share would readily be replaced by larger producers were it to cut back.) In a world as globally connected as ours, democracy can't end at a country's borders. Can we call it democracy, if the victims of fossil fuel extraction—the global poor who are losing their lands to rising seas, hurricanes, and drought—have no voice in the policies of countries like Norway (and Saudi Arabia, Russia, and so on) that are destroying their livelihoods?

Right now, most of us have little or no say over how our energy is produced. In fact, most of us don't even know. We pay private companies to keep gasoline, electricity, and natural gas flowing into our homes and cars, to supply food to our supermarkets, and to keep our economy humming. But we generally have no idea how and where the fuels are produced and extracted, how much money company executives make, or how much their shareholders are paid in dividends.

When we have the chance to find out, a lot of us don't like what we see. When I take people from my hometown of Salem, Massachusetts, to Colombia to visit the coal mine that until a decade ago supplied coal to our local power plant, they are horrified. They see the Indigenous and Afro-Colombian peasants displaced and made homeless by the massive operation, traditional livelihoods and farmlands destroyed, rivers dried up and diverted, and dust contaminating air, land, and water.

Around the United States and the world, frontline communities have mobilized to resist coal mines, pipelines, natural gas fracking operations, and oil drilling. But our energy system keeps us so distanced and ignorant, most of us just see the light switch in our kitchen or the pump at our local gas station.

In some ways, it is easier to remain ignorant. It's awfully convenient to have all of the energy and products we want at our fingertips, without having to think about how they got there.

But it's morally problematic, too. Most of us don't actually *want* to be complicit in human rights violations and environmental destruction. Not only are we, in the United States, privileged because of our access to way more than our fair share of the world's energy resources, we're privileged by not having to suffer the discomfort of knowing or seeing the real-world consequences of our consumption.

Like a just transition, energy democracy is a concept that can help to highlight common interests among consumers, workers, and frontline communities. Trade Unions for Energy Democracy (TUED), formed in 2013, is a coalition of twenty-nine unions and national and international labor federations. Its mission is "to advocate strongly for public direction and social ownership of energy at the local to the global levels, to assist in laying the foundations for durable and effective alliances between unions and other social movements." Energy democracy, writes TUED's Sean Sweeney, means going beyond the idea of a greener capitalism to address "the fundamental question of who owns and controls energy resources and for what reason energy is generated and used."[55] In concrete terms, this means reclaiming public control over privatized parts of the energy industry, creating real democratic control over publicly owned energy operations, and developing a new publicly owned, unionized, and democratic non-fossil-based energy system.[56]

Energy democracy advocates offer differing scenarios for how democratic control could work. Some emphasize small-scale, decentralized, locally controlled, community-based renewable energy systems. Others believe that economies of scale and existing infrastructure mean we should focus on democratizing large-scale energy projects. Unions caution that, given high levels of unionization in existing energy systems, we must be sure decentralization does not become part of the larger corporate attack on unions.[57]

Some in the Global South call for energy sovereignty, or energy justice, noting that far too often "democracy" there has been a veneer for colonial and corporate domination. The concept of energy democracy is based on a more radical vision than one limited to political, electoral, and representative systems; it calls for public, grassroots, bottom-up control over basic economic institutions and decisions. Still, the terms *energy sovereignty* and *energy justice* more specifically emphasize the ways the Global South and the poor have been systematically exploited by our historical and current energy systems.[58]

The energy democracy, energy sovereignty, or energy justice movement seeks to create alternatives while directly challenging the institutions that control our energy system and the regulatory apparatus that sustains it. Most energy democracy projects focus on the distribution of energy in the form of electricity rather than on extraction and production and its uses in other sectors. But the principles of popular democracy and transparency also shape movements in communities affected by extraction; they could be expanded to different economic sectors that use energy. Because of the fundamental role of energy in our economy, energy democracy relates deeply to the struggle for social, racial, and economic justice.

CONCLUSION: SHOULD SOCIAL, RACIAL, AND ECONOMIC JUSTICE ISSUES BE LINKED TO THE FIGHT TO STOP CLIMATE CHANGE?

Why does the issue of social, racial, and economic justice matter to countering climate change? Some have argued that addressing climate change is so urgent, we shouldn't complicate the fight by adding other issues into the mix. Others argue that, from a strategic perspective, we must include these issues in order to create the kind of broad-based social movement we need to confront the entrenched power of the fossil fuel industry.

Justice matters for its own sake, but I've tried to show here that social, racial, and economic injustice are tightly bound with the history and institutions that have led us to the brink of climate disaster. Our world continues to be shaped by ideologies and practices of progress rooted in Europe's colonial expansion and exploitation of the resources and labor of people of color in Africa, Asia, and the Americas. Europeans destroyed traditional lifeways and displaced, dispossessed, and enslaved people of color in their drive to build a new industrialized world based on ever-intensifying extraction of the planet's resources.

That's not just ancient history. It's the very foundation of the world we live in today. Elites and corporations worldwide have created unimaginable wealth and luxury for themselves and for a growing global middle class. Even before COVID-19 made things much worse, 690 million of the world's people—8.9 percent—didn't have enough to eat. In Africa, 250 million, almost 20 percent of the continent's population, suffered from hunger; another 381 million of the world's hungry lived in Asia, and 48 million were in Latin America. If we include those who faced moderate food insecurity, which means they lacked regular access to nutritious and sufficient food, we're looking at two billion people—a quarter of the world's population. The pandemic and associated lockdowns further devastated those impoverished by the global economic system.[59]

If we erase history, we might conclude the world's poor, conflict-ridden regions just happen to be inhabited mostly by people of color. But it's our global social and economic system that places people of color at risk. Historical, systemic exploitation created a wealthy Global North and an impoverished Global South. In the United States, it has concentrated people of color in jobs with the lowest pay and the most dangerous conditions, in unstable housing, and in neighborhoods with few resources.

Scholars Barbara Fields and Karen Fields have cogently pushed for expanding our critique of racism and our understanding of racial justice by challenging the explanation that people are oppressed or discriminated against because of the color of their skin. That's a false logic that ignores the structural causes of racial inequality.[60]

It's not skin color that makes some people suffer more from coronavirus, or hunger, or climate change than others. The roots of racial, social, and economic exploitation are in the very system that has led us to the climate crisis. Those who created the crisis also have the resources to protect themselves from its worst ravages and even to use it to enhance their wealth and power.

One reason to link the fight against climate change with the fight for social, racial, and economic justice is strategic. "To break the power of the reigning elite and impose public priorities on the economy, we need to build a mass coalition of ordinary people," argues an important recent analysis. "Rebuilding public power will require tackling the inequalities and divisions that capitalism sows."[61] We need to bring issues together because we need to bring people together to build the power necessary to create change.

But the reasons go deeper than just strategy. Our global economic system is the result of five hundred years of colonialism, extraction, production, profit, and economic growth. It's stolen the land, labor, and resources of the colonized world, creating a global racial hierarchy. It's created fantastic wealth and power for the few, a high-consuming lifestyle for a significant and growing middle class, dispossession, exploitation, and poverty for many, and a climate crisis for all of us. We need to understand the system in order to change it.

CHAPTER 5

BROADENING THE LENS

This section looks at debates over the years about the carrying capacity of the earth and the relationship of population, economics, and environment. Is there a limit to how many people the earth, or an individual country, can sustain? Does more technology allow the planet to provide for a larger population, or does it cause more environmental destruction?

It also tackles some of the questions that underlie many of the issues we've discussed in earlier sections. How are fossil fuels built into the very core of our economy? What kind of transformations can we make that allow the world population to enjoy the benefits of industrialization and technology in an equitable and ecologically sustainable way? In this section, we confront some of our beliefs and assumptions about the current inequitable, unsustainable realities and suggest ways of envisioning and bringing about the radical changes we need for a just and sustainable future.

IS POPULATION GROWTH THE ROOT OF THE PROBLEM?

Philosophers, scientists, and others have been trying to calculate the earth's carrying capacity, or the number of people the planet can sustain, for centuries. It's not an easy question to answer, because a lot of variables can change the equation. Technological change increased the planet's ability to sustain larger numbers of people—by making it possible to produce more food, for example. But it also increased people's environmental footprint and our pressure on planetary boundaries.

The debate dates back, at least, to English scholar Thomas Malthus, who in 1798 proposed that the human population would grow faster than its ability to produce enough food and exceed the planet's "carrying capacity." For Malthus, the process was cyclical and inevitable: because the land's ability to produce food was finite, population increase would eventually result in famine, thus reducing the population back to a sustainable level. Malthus's theory is also associated with politics; he concluded, in industrializing England, that any social programs or services that helped the poor would only make things worse, by allowing them to keep reproducing. Instead, his theory went, we should allow nature to take its course and eliminate overpopulation through famine.

Advances in technology and industry appeared to prove Malthus wrong about the limits of agriculture. Technological improvements, from better and mechanized plows to the new types of seeds, pesticides, and fertilizers of the Green Revolution in the 1940s, greatly increased the capacity to produce food on a given plot of land. The planet could support billions more people than Malthus believed. Techno-optimists came to believe that technology could be mobilized to solve other pressures on the planet's boundaries in the same way.

Both because of Malthus's cruel politics, and because his predictions failed to take technological progress into account, his ideas fell out of favor. They reappeared in a different way in the late twentieth century. World War II and the advent of nuclear weapons revealed the horrifyingly destructive potential of technology and unsettled even the most committed techno-optimists. Two books written in 1948, William Vogt's *Road to Survival* and Fairfield Osborn's *Our Plundered Planet*, revived Malthusian concerns about the impact of the world's growing population on its natural resources. Rachel Carson published *Silent Spring* in 1962, calling public attention to the hidden environmental disaster being created by the growing use of the pesticide DDT, an essential ingredient in the Green Revolution's agricultural expansion.

Six years later, Paul and Anne Ehrlich's *The Population Bomb* began with the dire warning that "the battle to feed all of humanity is over." They predicted that hundreds of millions of people would starve in the following decades. Their ideas, like Malthus's, were picked up by those who advocated and implemented repressive policies against the poor and people of color to try to "control" population growth. And like with Malthus, their scare tactics backfired. When their worst-case scenarios failed to materialize, many dismissed their arguments. (They were also strongly challenged by organizations like Food First, whose 1970s publications showed that injustice, not scarcity, was the cause of hunger.)

So what is the relationship between population size and climate chaos?

Total carbon emissions, per capita carbon emissions, and population growth all show a dramatic change around the middle of the twentieth century. Yes, carbon emissions began to rise with the Industrial Revolution at the beginning of the nineteenth century, but until the middle of the twentieth, the rise was very slow. After World War II, both world population and per capita energy consumption shot up—as did total energy consumption and carbon emissions.

Population and energy use didn't go up equally around the world, though. Per capita energy consumption and emissions rose in the wealthy countries and among the growing middle class of some poor and very large countries, like China and India. Population, however, grew fastest in the poorest countries and among the poorest people.

In the United States, the total fertility rate currently hovers around 1.7 births per woman, below replacement rate. In fact, US fertility has been below replacement rate since 1971. National Public Radio reported the news gloomily, quoting analysts who said things like "It's a national problem . . . The birth rate is a barometer of despair . . ."[1]

In the European Union, the birth rate is even lower, just under 1.6 in 2017; in Japan, it is less than 1.5. Only in sub-Saharan Africa, the poorest region in the world, are rates of between 4 and 5 births per woman common.[2]

Falling fertility rates don't necessarily mean population decline, at least not right away. If the number of young women entering childbearing years is increasing, the number of births can increase even as fertility rates decrease. As people live longer, the population can grow because births exceed deaths. (The counterpart here is also true: a high birth rate doesn't necessarily mean that the population will grow, if it's accompanied by a high death rate.) And migration can also lead to population growth or shrinkage. In the United States, the population has continued to grow, from 152 million in 1950 to 329 million in 2019, even though the birth rate started to decline around 1960. The US Census Bureau's US World and Population Clock in mid-2021 listed one birth every eight seconds, one death every twelve seconds, and one international migrant every 645 seconds, coming to a net gain of one person every 25 seconds.[3]

The world population is likewise increasing. The poorer a region is, the faster its population has grown. Africa's population rose from 244 million in 1955 to 1.3 billion in 2019; there were increases during the same years in Europe (from 577 million to 743 million), Asia (from 538 million to 1.9 billion), and Latin America (from 193 million to 658 million). The world population increased from 2.8 billion to 7.7 billion between 1955 and 2019. Studies suggest that the population will peak around 2064, at approximately 9.7 billion, before shifting to a decline.[4]

To understand why the world population is growing so fast, why the rate of growth varies so much around the world, and why this growth is likely to reverse course soon, let's look at two clear trends. The first is that the world population started rising a lot faster in the twentieth century; the second is that this rise seems to vary with income: the poorer the country, the faster its population is growing. Demographers understand pretty well by now why this is the case. It has to do with what they call the demographic transition.

During most of human history, both birth rates and death rates were very high. Not until after 1700 did the world start to see significant population growth, and not until 1800 did the world population reach a billion. Since then, the global growth rate has accelerated nonstop—but unevenly.

Falling death rates were the first stage of the demographic transition. Improvements in sanitation, healthcare, and, eventually, vaccinations started to bring down death rates, especially among infants and young children. So in the first stage of the demographic transition, birth rates didn't change, but the population began to grow because of falling death rates. That meant more infants surviving childhood, more children growing up to reproduce, and more people living into old age. This transition began around the late 1700s with the Industrial Revolution in Europe and the United States, and it happened gradually, as scientific knowledge, public health measures, and infrastructure developed. Furthermore, at the same time the populations of Europe and the United States began to grow, they exported a lot of people, which cushioned the process. Europeans came to the Americas, and the United States expanded westward.

In the Third World, access to health measures that spurred the transition came much later and much more quickly, so the rise in population was correspondingly faster. In sub-Saharan Africa, the child mortality rate in 1900 was about 500 per 1,000; only half of the babies born survived past their fifth birthday. By 2000, child mortality had declined to 150 per 1,000, still much higher than the rate in high-income countries but much lower than a century earlier. As Third World populations grew, other countries closed their doors, so there was nowhere for them to go. And climate change was making their homelands less habitable. For all these reasons, population growth was harder to adapt to.

In the second stage of the demographic transition, which began in Europe and the United States in the 1900s, birth rates begin to go down. As more children survived childhood, parents tended to feel less pressured to keep having children. As more healthcare, and especially family planning techniques, became available, women gained a greater capacity to control their fertility. As more opportunities, including education, became available to women, the desire to bear large numbers of children likewise declined. As people moved from rural to urban areas—another trend that has accompanied the global changes described here—children became an economic liability rather than a benefit. In the village, on the farm, young children quickly became another pair of working arms. In the city, those children had to be fed, clothed, and sent to school, so raising them became increasingly costly. This second stage has clearly begun in much of the Third World, especially in middle-income countries. Birth rates between 1960 and 2005 declined by 61 percent in Asia and Latin America, and by 37 percent in Africa.[5] But some believe it's not happening fast enough.

Some countries, like China, have tried to implement drastic policy measures to lower their birth rates. Others, including the United States, have used eugenics and involuntary sterilization to try to reduce unwanted populations. These kinds of measures are violations of human rights.

The most effective and comprehensive ways to slow population growth turn out to enhance, rather than to violate, human rights. They're the same changes that brought about the second stage of the demographic transition in today's wealthy countries: access to healthcare, education, and opportunities, especially for girls and women. Of course, these are worthy goals regardless of impact on birth rates. But if we are concerned about limiting population growth, more equality is the way to get there.

Reducing population may seem like an obvious step towards reducing the human impact on the climate. Every person creates an environmental impact. But numbers don't tell the whole story, because the impact an individual has on emissions varies greatly according to where they are and who they are—basically, according to how rich they are and how much they consume. If the goal is to reduce carbon emissions by lowering population, we should focus on lowering the population of rich people.

But rich people already have low birth rates, although with access to safe and healthy environments and to healthcare, they have low death rates too. Yet the rich also consume far more than their share of the planet's resources. More equality—reducing overconsumption by the world's wealthiest and redistributing access more fairly—is a more climate-effective goal than trying to reduce numbers that are going down anyway.

The advanced capitalist countries of the Global North account for 17 percent of the world population and 77 percent of CO_2 emissions since 1850. Global emissions increased by a factor of 655 between 1820 and 2010, while the planet's population increased by a factor of only 6.6. In many cases high population growth is *negatively* associated with high emissions growth. This makes sense. Poor countries may have higher population growth than rich countries, but each person consumes much less.[6]

The planet's human population is indeed putting pressure on the planetary boundaries identified by the Stockholm Resilience Centre. But much of humanity lives in ways that contribute very little to this pressure. Rather than worrying about the sheer number of humans on the planet, which is headed towards stabilization and decline anyway, we should focus on the over-consumers—or on the corporations, governments, laws, systems, and structures that are responsible for our skyrocketing emissions, use of resources, and creation of waste.

IS IMMIGRATION BAD FOR THE ENVIRONMENT?

The environmental movement has historically viewed humans somewhat dubiously. It has flirted with attempts at population control. One offshoot of this debate relates to the role of immigration in climate change. Some US environmentalists and environmental organizations have opposed immigration on environmental grounds. Preserving the environment by restricting immigration is compatible with many conservation organizations' goal of setting aside and protecting specific wilderness areas, often by expelling the Indigenous population. Restricting immigration is like designating the entire country as a protected area.

Conservationism, nativist racism, and immigration restrictionism intertwined.[7] Mainstream US environmental organizations founded in the late nineteenth and early twentieth centuries like the Sierra Club openly promoted racism and immigration restriction. Only very recently have these organizations begun to challenge a reflexive anti-immigrant position. Even some more radical environmentalists have been slow to disentangle their thought from some of the nativist and racist roots of the movement.

The debate simmered in the Sierra Club for decades, and in 1998, amid a growing national debate on immigration, the organization adopted a formal policy that it would not take a position on the topic. In 2004, an anti-immigrant contingent sought to take over the organization's board of directors, and the controversy surged into the public sphere. Bill McKibben, 350.org founder, responded in an article in the environmental journal *Grist* entitled "Does It Make Sense for Environmentalists to Want to Limit Immigration?" The Sierra Club faction that wants to limit immigration, he argued, should not be dismissed as racist:

> This is an important question. Or at least a subset of an important question: Does the size of America's population matter? It's worth trying to figure out the answer, because the U.S. Census Bureau estimates that at our present pace we'll grow from our current size of almost 300 million people to nearly half a billion by sometime later this century. Immigration accounts for most of that increase, both directly and because immigrants, at least for a generation, tend to have larger families.

> Half a billion is a lot of people. Try to imagine almost twice as many of us squeezing into the same towns, parks, schools, hospitals, roads. But more to the point, it's a lot of Americans. When people think about population, they tend to think about Africa and Asia and Latin America, where growth rates are much higher. And clearly Malians or Bengalis can damage their own prospects with rapid population growth. But they can't do much damage to, say, the upper atmosphere—they simply don't use enough stuff. If you're worried about shredding the global environment, the prospect of twice as many world-champion super-consumer Americans has got to worry you . . .
>
> At which point, of course, you could conclude that our job is to cut American consumerism . . . Which of course it is . . . But you have to be optimistic indeed to anticipate that happening so quickly that another couple of hundred million Americans wouldn't make a difference. So I think that the immigration-limiters running for the Sierra Club board have a reasonable point.[8]

To their credit, both McKibben and the Sierra Club have since renounced their anti-immigrant stances, influenced in part by the growing immigrant rights and environmental justice movements. By 2018, 350.org was supporting "the right of people to migrate." The Sierra Club too was outspoken in its opposition to the border wall and other anti-immigrant policies from the Trump administration. McKibben has also pointed out that many immigrants are climate refugees, forced from their homes by drought and other climate-induced disasters.[9] President Trump's anti-immigrant rhetoric and policies seemed to have pushed many in the environmental movement to think more deeply about the racist implications of immigration restrictionism.

But the positions articulated several decades ago continue to resonate in sectors of the environmental movement, putting an environmental or climate twist on nationalist and nativist anti-immigrant arguments. Don't immigrants place more pressure on already fragile environments, use resources, consume *more* in the United States than they did in their countries of origin, thus contributing more in emissions? And, with their remittances (the money they send home to families), don't they also contribute to raising consumption levels in their home countries?

Immigrants are, of course, a diverse group. While we hear a lot about the US-Mexico border, it's only a small piece of the larger picture of immigration. Mexicans have not been the largest group of immigrants entering the United States since 2013. Since then, India and China have vied for the top spot, with Mexico in third place. Smaller but still significant numbers come, in descending order, from the Philippines, El Salvador, Vietnam, Cuba, the Dominican Republic, Guatemala, and Korea. Together, these ten countries accounted for 60 percent of the US immigrant population. Close to half of immigrants who entered since 2013 hold a bachelor's degree or higher, as compared to just over 30 percent of all US-born adults. Overall, those foreign-born are likely to fall at the higher and lower ends of the spectrum in income and education than are those native-born, but their diversity in many ways mirrors the diversity of those born in the United States.

People of Mexican origin still constitute the largest population of US immigrants, at 24 percent, because of larger numbers in the past. Forty-four percent of the immigrant population identifies as Hispanic or Latino. Like most debates about immigration, debates among environmentalists tend to focus on this sector of immigrants.[10]

Many immigrants come from countries with higher birth rates, and it's true that immigrants tend to have more children than native-born Americans. But birth rates decline quickly with immigration, and even more in the second generation. So overall immigration, like rural-to-urban migration, contributes to lower birth rates.[11]

It's true that all people in the United States, including immigrants, consume and emit more than those in most other countries. It's also true that remittance flows tend to increase consumption of both high-carbon (vehicles, fertilizer, appliances, and travel) and low-carbon goods (education and healthcare) in poor countries. But these facts are more about economic development and the global economic system than about immigration per se. Rather than decry the increased consumption that comes with immigration, we might ask "Do poor people have the right to a bigger piece of the global pie—including the global carbon budget?"

Climate change contributes to migration in a multitude of ways, since increasing temperatures contribute to droughts, extreme weather events, fires, and rising sea levels, all of which can undermine agricultural production, render communities uninhabitable, and exacerbate social and land conflicts. The rural and urban poor are most vulnerable to these impacts. They comprise the majority of the new category of climate refugees.

As long as our global economic system creates islands of affluence amidst seas of impoverishment, and that affluence drives a changing climate turning those seas of impoverishment into disaster zones that prompt the flight of millions of refugees, immigration must be understood as an essential climate issue. Social justice requires not only that we provide safe harbor for climate refugees, but that we challenge the system that creates climate refugees in the first place.

WHAT IS ECONOMIC GROWTH? HOW IMPORTANT IS IT, AND HOW DOES IT AFFECT THE ENVIRONMENT?

Economic growth means an increase in production and consumption. Most economists, and the general public, think of economic growth as an indisputably positive thing. They connect it with more jobs, more availability of products, more consumption, and rising living standards.

Economic contraction, on the other hand, brings recession and depression. Demand shrinks, factories cut back production, people lose jobs, and the economy spirals downward, bringing unemployment and hunger. During the Great Depression of the 1930s, and to a lesser extent, again during the recession of 2008, the government stepped in to create demand by increasing its own spending and hiring, spurring production and jobs, thus consumption (because the government buys things itself, and because people start to make enough money to buy goods and services again), thus more production and jobs. Economists talk about the "health of the economy" and "national wealth" as depending on continuous growth.

But the economic growth model ignores some important issues. Gross domestic product (GDP)—the sum of goods and services produced in the national economy and the most common measure of economic growth (or contraction)—assumes that more is always better. But this is not necessarily the case. More consumption of food may bring better health and well-being, up to a point, but after basic nutritional needs are met, more consumption of food can lead to obesity and diabetes. More consumption of cigarettes or nuclear weapons can harm, rather than improve, human well-being. A human-made or natural disaster can add to the GDP, because it creates spending and jobs to clean it up or rebuild. GDP treats all consumption as a benefit, ignoring the many ways in which this may not be true.

Economic growth also fails to measure other aspects of well-being. Humans have social, emotional, cultural, and other needs as well as material ones. People need fresh air and clean water; they need family and community ties; they need meaningful engagement with the natural world, other humans, and social and cultural activities.

Economic growth, social justice, and planetary boundaries

Economic growth raises serious issues in terms of both social justice and the environment. GDP gives an overall number; it doesn't reveal how economic growth is distributed. If a few people get very rich and spend a lot of money, the GDP can go up while the poor stay poor. And GDP ignores the environmental costs of economic growth. More cars might seem like an improvement in standard of living, but they cause more pollution.

In some mythical ideal economy, economic growth might mean that everyone gets to consume more and benefit equally. (If we accept the idea that increased consumption is always a good thing.) But in the real world, economic growth often benefits the few to the detriment of the many. For example, take the coffee export boom in Central America at the end of the nineteenth century. Central American economies grew, and plantation owners, governments, and exporters made a lot of money. US importers made money, and US consumers got a lot of coffee. But for Central American small farmers, many of them Indigenous, the coffee export boom meant being forced off their communal lands and into virtual slavery and impoverishment on the coffee plantations. El Salvador or Guatemala as a whole might have been making money, but poor people were becoming poorer.

Or take an example closer to home—the coal industry that came to central Appalachia around the same time. Small farmers and subsistence economies were destroyed as big coal companies took control. Workers were subjected to dangerous and toxic conditions. Mechanization after World War II left many miners jobless, and mountaintop removal mining further destroyed jobs, land, and communities. Even as Appalachia supplied fabulous wealth to the country as a whole, its people were impoverished. Some call this "growth without development," where the economy grows, but the growth doesn't make life better for the population.[12]

How, then, can we improve material conditions for the world's poor, many of whom continue to lack basic necessities like clean water, education, and healthcare, to say nothing of the kind of hyper-consumption we enjoy in the First World? Many of history's social revolutions, from the French Revolution of 1789 to the Nicaraguan Revolution of 1979, proposed redistribution: the Robin Hood method. If the rich consume too much and the poor get too little, the solution is to take from the rich and give to the poor.

The rich, naturally, don't like this solution very much. In fact, this issue underlies much of US reluctance to participate in international climate accords. The United States wants to make sure the accords don't harm its economy, force over-consumers to consume less, or undermine its economic growth and profits by imposing globally redistributive measures.

Those who oppose redistribution call instead for more economic growth. They argue the problem with growth thus far is there hasn't been enough of it. Increased production often does benefit ordinary people, at least in some ways. People in the United States, even poor people, live far more comfortably and consume far more than their ancestors two hundred years ago or most people in poor countries today. (The rich in the United States live far more comfortably and consume far more than the poor or even the debt-ridden middle class in the United States.)

Poor people worldwide desperately need access to basic material goods. But there are two problems with relying on economic growth to accomplish this. One, without redistributive policies, there is no guarantee the fruits of economic growth will actually trickle down to benefit the poor. And to the extent that it does, how long will it take, and how many will have to die in the process? And two, can sufficient increases in production and consumption be accomplished given the pressure we are already placing on our planet's systems? Either way, some form of redistribution is crucial.

CAN WE HAVE ECONOMIC GROWTH WITHOUT INCREASING EMISSIONS?

We've looked at the examples of plantation agriculture and mining, two essential components of our modern economy, which hint at the environmental costs of economic growth. Increased production almost inevitably means increased extraction of resources, deforestation, contamination of air, water, and land—and GHG emissions. If we consider growth in the global economy and in carbon emissions over the past two hundred years, we can see they have increased virtually in lockstep. In fact, despite the devastating effects of the 2008 recession on many of the world's people, it actually benefited the environment: carbon emissions dropped significantly because production slowed. (Emissions rebounded and then some, as the economy resurged in 2010.) Even more dramatically, the worldwide shutdowns during the first months of the COVID-19 pandemic cut emissions briefly by almost 20 percent, down to 2006 levels.[13]

Advocates of the economic growth model emphasize the social costs of these slowdowns, and argue that technological advances allow humanity to continue to produce just as much, or more, by decoupling GDP growth from increases in carbon emissions. Decoupling can be *absolute*: emissions go down while the economy grows, or *relative*: emissions can continue to rise but more slowly than GDP. The idea is that new, cleaner technologies can make growth greener.

There is some evidence that decoupling is, in fact, occurring. In the United States, emissions declined slightly from 2010 to 2017, while GDP grew. Some European countries have managed to achieve lower emissions along with economic growth in the last decade.

The most optimistic take on decoupling is that cleaner technologies are letting us produce more with lower emissions and/or we're shifting our economy towards more low-emissions sectors like education or cultural activities. Many progressive economists, and Green New Deal supporters, advocate policies to increase job-creating investments in green infrastructure and technology. They argue ongoing decoupling will enable us to lower emissions sufficiently while economic growth continues.[14] But the cases where decoupling has occurred contain some obvious, and some hidden, complexities that undermine the optimism.

The shift from coal to natural gas and decoupling

Much of the slowdown in emissions from the United States and Europe was due to a shift from coal to natural gas. While it's true natural gas creates lower emissions than coal, the shift is not a recipe for continuing to decrease emissions—it's more of a one-time drop. If we continue to produce more, using more natural gas, emissions will increase again. (Plus, as discussed earlier, methane emissions from natural gas often go uncounted.)

This is exactly what happened in the United States. Between 2010 and 2017, emissions dropped as natural gas replaced coal in the power sector. But economic growth and weather conditions (cold winter, hot summer) brought a 10-percent increase in natural gas use in 2018, reversing the trend and raising emissions again.[15]

The International Energy Agency (IEA) reported flat global emissions in 2019, suggesting optimistically it could represent "the definitive peak" of CO_2 emissions. Other analysts were not convinced. One study warned that "a sharp drop in coal generation was responsible for lower emissions in the United States and Europe—and that everywhere else the picture looked bleak." Only between 2014 and 2016 did global emissions appear to decouple from economic growth, and the divergence could be almost wholly attributed to the decline of coal in the United States and the EU. In fact, even as US power generation emissions declined, emissions continued to rise in transportation, industry, buildings, and other sectors. And the IEA figures measured only CO_2 emissions, omitting the growing methane emissions from natural gas.[16]

Growth in low-emission sectors

Another factor that can contribute to decoupling GDP growth from rising emissions is expansion of low-emission sectors of the economy. Growth in some low-emission sectors, like education, can have a clear positive impact on people's lives even though it doesn't entail greater production and consumption of material goods.

Much of the low-emission expansion in recent decades, though, has been in less socially useful sectors, like FIRE (finance, insurance, and real estate). FIRE includes profitable activities that speculate, invest, and transfer vast sums of money but don't employ many people, create many consumer or social goods, or contribute much to the standard of living. And FIRE sector transactions have been increasingly incorporated into GDP accounting.

Growth in the FIRE sector can in fact be detrimental to many ordinary people. Real estate speculation—the housing bubble—was a major factor behind the 2008 recession. If your rent goes up, or you pay interest on your student loan, you're contributing to the GDP without much benefit from it. From a social perspective, financialization contributes to growing inequality, making GDP even less relevant for measuring the well-being of the population.[17]

But like investment in more socially useful low-emission sectors, financialization can appear to contribute to decoupling. It makes it look like the economy is growing, when in fact it's just rich people moving their money around and making more money with it. This form of decoupling doesn't mean that we're producing more with less environmental harm—it means that we're measuring nonproductive activity and incorporating it into our GDP calculation.

The GDP and emissions embodied in trade

The flow of imports and exports affects the relationship between GDP and national-level emissions statistics. If fossil fuels are burnt elsewhere but profits and products come to the home country, GDP may grow while the emissions associated with that production are hidden. When Norway exports its petroleum, it gains in GDP but takes no responsibility for the emissions when that oil is burned. If the US exports low-emissions goods and imports high-emissions goods, it can "displace" its emissions while continuing to grow consumption and GDP.[18]

Two cases often cited as evidence that decoupling can occur are Sweden and the UK, but a closer look reveals economic growth without increased emissions in these countries has occurred partly through changing import and export patterns. One factor in both countries' emissions reduction, concluded a 2018 study, was the shifting of domestic production to lower-emissions goods and outsourcing the production of higher-emissions goods.[19] So, Sweden and the UK can claim lower emissions while consumption grows because the emissions from what they consume are now occurring in other countries. In fact, if we include emissions from shipping and factor in that carbon-intensive industries are frequently outsourced to countries with lower environmental standards, the cumulative effect of outsourcing may be to *increase* emissions worldwide.

These two cases represent a larger trend. In the early-twenty-first century, a full one-fourth of global emissions came from products exported from China and other Third World countries for consumption in the United States and Europe. Yet the emissions are counted towards the producing country's share. Since 2001, the year China entered the World Trade Organization and removed restrictions on foreign investment, 50 percent of its increases in emissions have come from the export sector. Between 1990 and 2008, emissions from "emerging economies" due to their production of exports to "advanced economies" quadrupled. In a single year (2011) export-oriented production in poor countries created 2.95 gigatons of CO_2 emissions. If we counted these emissions towards the consuming countries, we'd find that no decoupling at all had occurred: "developed countries have not recorded a decrease from 1990 levels but rather an increase." Another study in 2019 found the United States, Canada, and Australia all followed the UK pattern of outsourcing emissions.

A slightly more optimistic study reported that while part of decoupling in advanced economies is due to outsourcing dirty production, even when we measure consumption-based emissions, some degree of decoupling is occurring.[20]

So back to the original question: Is decoupling for real? Can we have economic growth without increasing carbon emissions? The answer seems to be maybe, sort of, a little bit—but not much. Certainly not enough, and not fast enough, to avert catastrophic climate change. Some of the apparent decoupling we've seen comes from the way the numbers are counted—including financialization as economic growth and not counting the emissions when we shift production abroad. While it's true that technology and policy can make our intensive energy use somewhat less damaging, it's also time to think about questioning our whole commitment to economic growth.

DO WE EVEN NEED GROWTH? WHAT IS DEGROWTH? IS IT A GOOD IDEA?

This book has hinted at the concept of degrowth from the beginning. We live on a finite planet, with finite resources. We in the First World already consume far more than our fair share and have contributed far more than our fair share to global GHG emissions. And we are already pushing our planet's boundaries on multiple fronts. How can we expect to keep consuming and producing more? How can we justify producing more luxuries for the rich when the global poor lack access to the basics of survival? Degrowth economics argues we should scale down our material economies to a level consistent with the planet's finite resources and manage this degrowth in a way that prioritizes human needs and enables human flourishing. Degrowth critiques capitalism as a system that relies on increasing production, consumption, and use of resources for the benefit of the few, at the expense of the many and the environment.

The idea more is not necessarily better has been around for a long time. For much of human history, most people have lived in traditional societies: tribal, hunting and gathering, or small agricultural villages. These types of societies valued redistribution over accumulation and tradition over constant innovation. Their cultures and religions developed ways to guarantee survival through reciprocity and continuity. Human relations with the natural world were often structured through sacred and collective beliefs.

Anthropologist Marshall Sahlins described hunters and gatherers as "the original affluent society" because they are able to easily satisfy all of their material needs with very little labor. They certainly subsist on much less than the denizens of industrial societies, but "the customary quota of consumables [is] culturally set at a modest point."[21]

According to Sahlins, much of what we know about hunting and gathering societies is distorted by two factors that accentuate their material deprivation. One, as industrial consumers, the anthropologists who have studied these societies bring their own consumerist biases to the project. They have assumed low levels of consumption translate into deprivation and want, even when confronted with abundant evidence that the people they are studying do not experience it that way. Second, today's hunters and gatherers have been pushed by colonizers into the most marginal territories of the planet: the deserts, the tropical forests, the Arctic. Many of them have experienced colonization and conquest in the form of forced removal, forced labor, epidemic disease, and exploitation. So anthropologists are studying the survivors of holocausts, while assuming that they represent "untouched" cultures.

Studies of peasant societies have come to comparable conclusions. Yes, peasants work much harder to fulfill their needs than do hunters and gatherers. And yes, most peasants live in complex relation to larger societies, landlords, or states that may tax them, compel their labor, or purchase their products. Yet studies from around the world have noted ways in which peasant religion and culture value continuity and reciprocity over innovation, accumulation, consumption, and growth. Peasant societies have evolved with the knowledge that novelty or even intensification of agricultural techniques is as likely to bring disaster as benefit.

James Scott called this the "moral economy of the peasant." "The critical problem of the peasant family," he writes, is "a secure subsistence." Peasant families "typically prefer to avoid economic disaster rather than take risks to maximize their average income," Scott adds. "Patterns of reciprocity, forced generosity, communal land, and work-sharing helped even out the inevitable troughs in a family's resources which might otherwise have thrown them below subsistence. The proven value of these techniques and social patterns is perhaps what has given peasants a Brechtian tenacity in the face of agronomists and social workers who come from the capital to improve them."[22]

Growth, then, is a uniquely modern and industrial idea. Historian Steven Stoll described the cultural changes that accompanied the Industrial Revolution: "Between the 1820s and the 1850s, a new kind of existence came into view, powered not by lumbering bodies but by gravity and coal. The fusion of philosophical idealism with innovations in mechanics released a soaring optimism. Go back to 1750, and everyone on earth lived nearly the same way—moving only as fast as a horse, pulling only as much as an ox, and preparing food, shelter, and clothing by hand. It was a biological old regime about to be overthrown."[23]

The concept of growth became even more popular after the Depression and World War II, as development economics proposed capitalist growth as the solution to Third World poverty. Growth and progress, now often termed "economic development," became the colonial powers' justification for continued intervention in the context of Cold War anti-communism. As the colonies fought for independence and often for the overthrow of capitalism, which had not worked out too well for their people, development specialists promised that foreign investment and export-oriented growth would be their ticket to prosperity. Growth would be an antidote to revolutionary projects based on redistribution, which the United States universally tarred as "communist."

The World Bank and the International Monetary Fund, two international institutions created at the Bretton Woods Conference in 1944, would fund and oversee Third World economic growth that would raise living standards and reduce poverty. Economist Walt Whitman Rostow proposed a linear process every country could follow to achieve "economic take-off" and enter the era of mass consumption. Rostow's *Stages of Economic Growth* popularized the view that all economies follow a single trajectory towards increasing growth—starting with traditional agriculture and gradually intensifying mechanization and manufacturing (and fossil fuel use) to reach a final stage of mass consumption. For Rostow, this was both a description of history and a road map for "underdeveloped" societies to follow. The United States and international institutions could push the process along in what Nick Cullather called "America's Cold War battle against poverty."[24]

Many Third World peasants responded as Scott described, with justified suspicion as development initiatives pushed projects benefiting foreigners, urban dwellers, and elites while undermining rural lifeways. When subsistence producers are forced into the wage economy, it may look statistically as if their lives are better because they are earning more money than they did before—even as they lose their land, their community institutions and ties, and their security.

A new generation of Third World intellectuals offered a different explanation for Third World poverty, casting doubt on development economics' assumptions. "Dependency" and "world systems" theorists argued that First World economic development was based on colonial exploitation of the resources of the Third World in a process that created the latter's poverty and "underdevelopment." In their attention to the extraction and depletion of their countries' resources, these critiques carried a strong environmental flavor.

Growth had its critics inside the First World as well. Sahlins's book, which referred to Stone Age people as "the original affluent society," was a play on US economist John Kenneth Galbraith's *The Affluent Society*, a critique of mass consumption published in 1958. Galbraith argued modern industrial society had created mechanisms for constantly increasing wants and needs. Production and consumption came to be valued for their own sake, far beyond basic human needs or social welfare. Capitalism created a treadmill of increasing production and consumption to keep people employed, thus enabling them to keep consuming, thus enabling factories to keep producing, thus enabling owners and investors to profit and keep investing. The only alternative within the system seemed to be economic recession or depression.

Galbraith suggested reorienting the economy towards fulfilling a broader conception of human needs and security. Rather than relying on ever-increasing production for its own sake, he explained, a strong public sector should guarantee basic needs like health, education, housing, transportation. After that, the aim of the economy should be to increase leisure and make work more palatable—to "maximize the rewards of all the hours of their days"—rather than to mindlessly increase production and consumption.[25]

Galbraith's critique of overconsumption was based primarily on its effects on the overconsuming society itself rather than on the environment or the rest of the world. Although marginalized from mainstream economic thought in the United States, his perspective was pursued by a few economists. Richard Easterlin noted that wants seemed to increase with prosperity. The more the rich accumulated, the unhappier the poor felt with their lot, even if they too were consuming more.[26] In *The Overspent American: Why We Want What We Don't Need*, *The Overworked American: The Unexpected Decline of Leisure*, and *Born to Buy: The Commercialized Child and the New Consumer Cult*, economist Juliet Schor explored the ways that consumption proved addictive and immiserating, crowding out more essential and satisfying human needs.

The early 1970s saw a wave of environmental critiques of economic growth, beginning with the international think tank Club of Rome and its 1972 book *The Limits to Growth*, which argued that humans could not continue to increase their use of finite resources on a finite planet. E. F. Schumacher wrote *Small Is Beautiful: Economics as If People Mattered* in 1973; Herman Daly wrote *Steady-State Economics* in 1977. André Gorz's *Ecology as Politics* (first published in French in 1975) became a foundational text for the new field of political ecology, which sought to analyze the political causes of environmental destruction. Socialism as it actually existed, Gorz argued, fell into the same trap as capitalism in believing that humanity's problems could be solved by increasing production, albeit with better distribution. Gorz married the cultural politics of the New Left, a socialist critique of capitalism, and an ecological challenge to economic growth.

Forty-five years later, Gorz's description of the consequences of a system that disregarded the limits of the natural world, today called planetary boundaries, rings prescient: "New diseases and new forms of dis-ease, maladjusted children (but maladjusted to what?), decreasing life expectancy, decreasing physical yields and economic pay-offs, and a decreasing quality of life despite increasing levels of material consumption." Gorz explained that we should "improve the conditions and the quality of life" by "refusing to produce socially those goods which are so expensive that they can never be available to all, or which are so cumbersome or polluting that their costs outweigh their benefits as soon as they become accessible to the majority."[27]

Ecological economics, formalized by Joan Martínez-Alier in his 1987 *Ecological Economics: Energy, Environment, and Society* and by the founding of the International Society for Ecological Economics and its journal *Ecological Economics* in 1989, seeks to highlight the role of the natural world in the economy. Ecological economics, Martínez-Alier explains, "is a critique of mainstream economics" based on two critical points: "we must view the economy physically, counting energy and material flows (in calories or joules, or in tons), and we must abandon the Gross Domestic Product, which mixes up production and destruction . . . When the industrial economy grows, ecosystems are destroyed."[28]

Mainstream economics ignores the ecological costs of economic activity, like the pollution, including carbon emissions, and the exhaustion of resources inherent in productive activities. Ecological economics tries to find ways to calculate and factor in environmental costs, which make it clear the model of economic growth has severe shortcomings. If we subtract the environmental costs of production, much of what economists have long considered "growth" is revealed as a statistical trick, a figment of vivid imaginations. What economists call productivity and growth actually involves a lot of destruction.

Mostly, though, economic critiques of consumer society and the concept of growth were lone voices in the wilderness. Mainstream economists continued to believe that the rising tide of growth would raise all ships, eliminating worldwide poverty and want. International development agencies, the World Bank, and pundits from Steven Pinker to Nicholas Kristof to Bill Gates continue to take a triumphalist view of economic growth. Growth will benefit the poor, the story goes, and the environmental costs can be managed with technology to make it clean and sustainable.

Early socialist analyses of capitalism emphasized the system's exploitation and inequality. But they didn't question the need for industrialization and economic growth. Socialist countries like China and the USSR in the twentieth century could be as ruthless as their capitalist counterparts in exploiting and dispossessing peasants to accumulate resources for economic development and growth. The theory, which was only partially successful, was that the fruits of growth, rather than going into the pockets of the capitalist class, would be redistributed through social services. By the end of the twentieth century, Marxist scholars began to highlight capitalism's exploitation of nature as well as of labor.[29] Twenty-first century socialist experiments in Latin America struggled to reconcile their need for increased production and the export economy to fund their ambitious social projects, and the social and environmental exploitation involved in their growth agendas.

The degrowth school of economics draws on earlier critiques of growth but did not come together as a discipline until the beginning of the twenty-first century among a small group of European scholars and activists. Its official founding dates to a series of international conferences that began in Paris in 2008. Degrowthers joined academic and political movements for ecological economics, political ecology, and environmental justice in Europe and Latin America. They found a home in the Institute of Environmental Science and Technology (ICTA) at the Autonomous University of Barcelona in 2010, which now offers the world's first master's degree in political ecology, degrowth, and environmental justice.

Degrowth proposes that the world's "developed" economies have far exceeded their fair share in use of the planet's resources. Growth is based on increasing the extraction and use of the earth's resources, which are finite. All production also creates waste, including our increasingly perilous levels of carbon emissions. Endless growth is a logical impossibility. And the bigger an economy is, the more environmental destruction its continued growth is going to cause.

Rather than exporting their model of economic growth, overdeveloped countries need to shrink their material economies. Scarce material resources should be utilized in a way that prioritizes fulfilling human needs rather than profit. To do this, we need more popular democracy, a vastly expanded public sector, and community-based forms of sharing and exchange outside the profit system.

Following dependency theory, degrowthers argue that for the planet's poor majority, economic growth for the few has meant dispossession, exploitation, land loss, and environmental destruction. Growth is fundamentally unjust "because it benefits from an unequal exchange of resources. . . . The energy and materials that fuel growth are extracted from commodity frontiers, often in indigenous or underdeveloped territories that suffer impacts of extraction. Waste and pollutants end up in marginalized territories, communities or neighborhoods of lower class or of different colour or ethnicity than the majority of the population." Poor people's struggles for social and human rights are often environmental struggles as well, Martínez-Alier and Ramachandra Guha argue. "The environmentalisms of the poor," they explain, "originate in social conflicts over access to and control over natural resources."[30]

Degrowth is more than just a critique of existing systems of growth and inequality; it's also a proposal-under-construction for a completely different form of economic organization, locally and globally. "The alternatives, projects and policies that signify a degrowth imaginary are essentially non-capitalist: they diminish the importance of core capitalist institutions of property, money, etc., replacing them with institutions imbued with different values and logics. Degrowth therefore signifies a transition beyond capitalism," wrote three of its leading proponents.[31]

"Beyond capitalism" economic proposals include local and grassroots institutions that operate outside the capitalist economy, along with a public social-welfare system aimed at fulfilling basic needs and reducing unnecessary work. You can get a glimpse of what such a system might look like by comparing European systems (especially as they functioned before the neoliberal assault of the late twentieth century)—with a strong social welfare state, shorter work hours, and longer vacations—to the United States. Countries like Spain, Portugal, and Italy have a much lower GDP per capita and much lower CO_2 emissions per capita than the United States. They also have guaranteed healthcare and more vacations—and longer life expectancies.[32] Degrowthers also point to much poorer countries like Costa

Rica and Sri Lanka, which have created robust health, education, and welfare systems that sustain high levels of well-being despite much lower consumption levels. Producing, consuming, and spending more doesn't necessarily make our lives better.

While capitalism prioritizes the productive economy and the potential for profit, degrowth highlights the relationship between profits, high resource use, and high emissions. (While high-profit transactions, like those in the FIRE sector, don't directly produce emissions, they still depend on fossil-based production or, like real estate bubbles, they turn into Ponzi schemes, that eventually collapse.)

In a degrowth scenario, low-fossil, low-profit sectors like the reproductive or care economy (childcare, education, elder care, land regeneration, and peasant agriculture) may expand through public investment or social provisioning (grassroots, community-based, and outside the market), but in a decommodified way, so they won't exactly look like "growth." Reproductive labor is the essential work that fulfills basic human needs, but it's devalued under capitalist, profit-oriented economics. Those who do the most essential work are among the lowest paid. When left to the capitalist market, the only way to make these sectors profitable is to serve only the rich, and/or squeeze workers. With a strong public commitment, these sectors can be great equalizers in society, but they are precisely what's being defunded and privatized.

Degrowth says we should work (for pay) and produce less, especially less socially and environmentally harmful things. The purpose of work should be to benefit (rather than harm) our community and our world. Under our current system, working less means losing benefits and security. But imagine if working less just meant fewer yachts and less transatlantic travel for your CEO, while you could enjoy healthcare and other benefits as well as more leisure and more opportunities for fulfilling activities. Degrowthers call this conviviality (which they take from philosopher Ivan Illich)—"personal creativity and collaboration . . . space for relationships, recognition, pleasure and generally living well" beyond "dependence on an industrial and consumerist system."[33]

Some have critiqued the field for focusing too much on how overdeveloped societies can "degrow" and not enough on potential paths for the world's poor whose problem is lack of resources, not overconsumption. Degrowth is fundamentally anti-colonial in its emphasis on the ways that growth in the Global North is based on the extraction of Global South resources. It proposes that overdeveloped societies must degrow in order to share the world's resources more equitably and allow the world's poor to consume more.[34] What this actually means as a path to development for poor countries is less clear.

In the Third World, where peasant economies are still vital and have been a strong source of resistance to extractivist development projects, degrowth resonates with concepts like *buen vivir* in Latin America, Gandhian economics or the "economics of permanence" in India, and the Ubuntu philosophy in Bantu-speaking Africa. These concepts look to local traditions that value human relations with the natural world and promote economies based on reciprocity and human needs. They challenge colonial—capitalist—economics because it exploits people and natural resources in the colonies, and they consider it an unworkable and undesirable model.[35]

Capitalists have long struggled to instill industrial discipline and consumer desires in a reluctant workforce, both in industrial cores and in colonial peripheries. Still today, whether in postindustrial societies like the United States and Europe or in rural peasant societies in the Third World and beyond, dreams of an economic system that prioritizes human needs and the rights of nature over profit and consumption continue to percolate. "The end of growth will not mean the end of progress," Stoll writes, "to the extent that we can redefine progress as consisting of something other than accumulation."[36] This is exactly what degrowthers in the Global North and South are attempting to do.

Beyond GDP

Most contemporary economists assume that GDP serves as a proxy for the well-being of the population, even though Simon Kuznets, who developed the measurement, made it clear that economic welfare was more complicated than that. Others have suggested more meaningful ways to measure how an economy is working. Herman Daly and John Cobb proposed the Index of Sustainable Economic Welfare (ISEW) in 1989, now generally updated to the Genuine Progress Indicator (GPI). Instead of counting up what's bought and sold, they explained, we could measure economic activity by its impact on human welfare. The ISEW adds in unpaid labor like childcare and volunteer work, which contribute to well-being, and subtracts resource depletion, which harms well-being. It further subtracts "defensive" economic activity that exists only to remedy problems caused by the economy itself, like cleaning up oil spills. And the ISEW counts inequality as a negative factor.[37]

Even the United Nations has recognized the limitations of GDP. In 1990, the UN proposed an alternative, the Human Development Index (HDI), which includes per capita income but demotes it to one of several factors in its calculation of "development." The HDI places equal importance on health, measured by infant mortality and life span, and access to education, acknowledging that GDP alone may tell us very little about the well-being of the population.

Yet another way of measuring well-being is subjective. Since 2011, the United Nations has compiled a National Happiness Report based on how people in different countries perceive their own levels of happiness and well-being.[38] The surveys the UN carried out back up the fundamental arguments made against economic growth and in favor of measures like the ISEW and the GPI. Once basic needs are met, what most contributes to human satisfaction is not more stuff, but rather close relationships, control over one's life decisions, social trust, and sense of equality. These factors correspond closely to what degrowthers call conviviality.

Some poor societies can achieve far better outcomes in well-being than the overdeveloped countries. Jason Hickel flags Costa Rica as "one of the most efficient economies on earth: it achieves high standards of living with low GDP and minimal pressure on the environment." It does so through its strong social policies, including universal access to healthcare, education, and social security.[39]

Moving away from the GDP helps us understand how degrowth doesn't necessarily mean belt-tightening. Producing and consuming less can mean more of the things that truly matter for a good life.

Some critics argue that degrowthers focus too much on lowering the GDP. But that's not precisely what degrowthers advocate. Rather, they are interested in how to create new systems and institutions that allow people to thrive independently of GDP. In the current system, a falling GDP brings economic crises like those of 2008 or 2020—or 1929—where the economy crashes, businesses close, and people lose their jobs, their homes, and their savings.

But it doesn't have to be that way. "Because degrowthers are liberated from the illusion that growth will carry on forever," writes Jason Hickel, "we openly call for an economy that does not require growth. We call for policies like a shorter working week, a job guarantee, debt-free currency, fairer wages, progressive taxation, universal social services, and so on, so that we can scale down economic activity in a stable manner, while at the same time facilitating human flourishing."[40]

So what is a reasonable GDP that can sustain social welfare within planetary boundaries?

Economist Kate Raworth proposed the concept of "doughnut economics" to help visualize how a society can fulfill basic human needs and keep consumption of resources within the nine planetary boundaries identified by the Stockholm Resilience Centre. The outer rim of the doughnut depicts the planetary boundaries, the limits of how much we can consume and emit before we push ourselves into ecological catastrophe. The middle level—the doughnut itself—represents the fulfillment of basic human needs, based on the UN's 2015 Sustainable Development Goals. Some of these needs are material: housing, food, water, and energy. Others are social: political voice, networks, peace and justice, social equity, and gender equality. Some, like health, income and work, and education, have both material and social dimensions. The hole in the middle of the doughnut represents an economy that fails to fulfill basic needs. The doughnut, then, illustrates "an environmentally safe and socially just space in which humanity can thrive."[41]

Energy journalist David Roberts suggests the doughnut can help us transcend the apparent contradiction between the need for growth to meet human needs and the disastrous environmental consequences it brings. Rather than judging a policy by its impact on growth, he suggests, "Good policy will raise up social indicators into the doughnut without pushing ecological indicators out of it, or vice versa. Bad policy will push one side at the expense of the other."[42] In other words, rich countries need to degrow their economies (and lower production and consumption) to fall within planetary boundaries, while poor countries may need to grow (that is, increase production and consumption) in order to fulfill basic needs.

Neither Raworth, who says she agrees with the ideas of the degrowth movement though she does not find the term politically useful, nor the degrowthers advocate shrinking the world's poorest economies. Degrowth focuses on how to shrink overconsuming economies, which are already outside the doughnut, in ways that will privilege basic needs and human flourishing beyond the demands of the market.[43]

An international group of scholars tried to quantify Raworth's vision in the recent article "A Good Life for All Within

Planetary Boundaries." The authors conclude that right now, no country in the world manages to meet its inhabitants' basic needs without transgressing these boundaries. Basic needs, including access to food, sanitation, and electricity, and the elimination of extreme poverty could be met everywhere while remaining inside the doughnut's safe zone. But this would require radically reducing resource use among the world's wealthiest to allow the poor to use more.[44]

IS DEGROWTH COMPATIBLE WITH THE GREEN NEW DEAL?

Because the Green New Deal (GND) is a broad concept right now, it encompasses a variety of different specific proposals. Some formulations of the GND, and some offshoots, like President Biden's plans, promote the idea of green growth, with a focus on new job creation and technology development. The green growth approach relies on a heavy dose of techno-optimism that remains unsubstantiated in the real world: thus far, our clean(er) technologies haven't made a dent in global emissions. Green growth avoids confronting the resource use and pollution created by all forms of production, including that of so-called green or renewable energy. Even as green growth often advocates some forms of redistribution domestically, increasing corporate taxes and public services, for example, US versions tend to assume the country can continue its high levels of consumption and corporations their high levels of profits.

Other GND supporters acknowledge the impossibility of ongoing growth in the overdeveloped economies and look for ways to slow down, work less, and expand nonmarket sectors rather than material consumption for what we seek as a good life. The degrowth, decolonial GND perspective sees redistribution as a global issue, emphasizing the need for reparations and sharing fairly the planet's finite resources, including our rapidly shrinking carbon budget.

The Green New Deal for Europe (GNDE) pushes the conversation further than most of the debates in the United States, emphasizing the need to degrow overconsuming economies, explicitly acknowledging planetary boundaries, and rejecting green growth. The authors contrast the GNDE's public works program with the original New Deal, which based its definition of economic recovery on restoring industrial growth. The GNDE instead aims to bring Europe's economy down to a globally sustainable level, based on the principle that the planet's resources, including the carbon budget, must be shared equitably.

The GNDE calls for "rapid transformation to an economy that respects planetary boundaries," reducing energy demand "by scaling down material production and throughput . . . Shifting income and welfare creation from industrial production to social and environmental reproduction: maintenance, recycling, repair, and restoration of environmental and infrastructural resources, as well as education, culture and care—for both people and planet."[45] All of this comes straight out of the degrowth philosophy, and indeed some of degrowth's best-known proponents (Giorgos Kallis, Giacomo D'Alisa, and Jason Hickel) are among its authors.

The GNDE's jobs guarantee recognizes that a decrease in material production and the elimination or reduction of whole sectors of the economy (beginning with fossil fuels) will require a decrease in labor. It thus proposes reduced working hours at increased pay to allow workers "the flexibility and security to choose the right balance of work and life." Reducing the private sector while greatly expanding the public sector will shift society's overall priorities from short-term profit and executive and shareholder compensation to the public good. Job creation would include compensation for what is currently unpaid care work and environmental restoration work. The goal is to "decouple social progress from environmental breakdown" and move the economy "away from extraction and consumption and towards regeneration and other socially valuable activities."[46]

Both the Green New Deal and degrowth reject the neoliberal economic model of privatization and austerity and advocate public investment to steer the economy towards meeting basic human needs. Even some green growthers acknowledge that slowing the economy down, and producing and consuming less (especially for the planet's 1 percent or even 10 percent, who are doing most of the consuming) can make our lives better, getting us off the growth treadmill and leaving more time to enjoy family, nature, education, slow travel, and other activities outside the market economy. Green growthers emphasize the potential of new technologies and especially green energy more than degrowthers. Degrowthers say only if the planet's over-consumers reduce can these new technologies be distributed fairly and keep us within the planet's carbon budget and other planetary boundaries.[47]

Most degrowthers reject market-based policies like carbon pricing that turn emissions into yet another commodity. Some see a carbon tax or cap and dividend as a potential tool towards a degrowth economy.[48] A degrowth approach liberates us to explore a wide variety of specific climate policy options. The main argument industry and its friends in Congress raise against policies like carbon taxes, hard caps, and strong regulation of fossil fuel extraction and fossil-consuming sectors like industry, transportation, and agriculture, is that such taxes or regulation would undercut economic growth. Degrowthers argue for the overgrown, overconsuming economies of the world that would be a good thing. If we can manage degrowth by guaranteeing basic public services, shortening the work week, and fostering low-carbon sectors like education, the arts, and urban and rural green spaces, life in our de-grown economy could be more equal, more leisurely, and more meaningful than the insecure work-and-spend rat race we live in today. And we can acknowledge that the formerly colonized, the global poor, have the same rights and needs as those who for too long have claimed their resources.

ARE WE MAKING PROGRESS?

"Nearly everything we understand about global warming was understood in 1979," Nathaniel Rich tells us on the first page of

Losing Earth: A Recent History. "It was, if anything, better understood." Better understood, because climate denialists had not yet intervened massively to attempt to undermine public confidence in the science. "The 'greenhouse effect'—a metaphor dating to the early twentieth century—was ancient history, described in any intro-to-biology textbook."[49]

The first World Climate Conference met in Geneva in 1979 and sounded the alarm. Over the next decade, we had an "excellent chance" of slowing carbon emissions to avert catastrophic climate change. In 1989, "the world's major powers came within several signatures of endorsing a binding framework to reduce carbon emissions—far closer than we've come since." Moreover, "during that decade the obstacles we blame for our current inaction had yet to emerge."[50] The fossil fuel industry had not yet mobilized to challenge climate science. Republicans and Democrats alike saw climate policy as a nonpartisan issue.

Yet when the UN's Intergovernmental Panel on Climate Change (IPCC) met in the Netherlands in 1989 to agree on a global treaty with enforceable targets, the United States led outliers from Britain, Japan, and the USSR in refusing any commitments to freeze emissions. "The final statement noted only that 'many' nations supported stabilizing emissions—but it did not indicate which nations, or at what level, or by what deadline. And with that, a decade of excruciating, painful, exhilarating progress turned to air." In fact, "more carbon has been released into the atmosphere since November 7, 1989, the final day of the conference, than in the entire history of civilization preceding it."[51]

Years later, the US delegate insisted that the conference was just a chance for posturing, because no country was really willing "to make hard commitments that would cost their nations serious resources." As one European delegate who supported a "binding" treaty remarked, "after all, it's only a piece of paper and they don't put you in jail if you don't actually do it."[52]

In fact, "even some of the nations that have advocated most aggressively for climate policy—among them the Netherlands,

Canada, Denmark, and Australia—have failed to honor their own commitments" to the 2015 Paris Agreement. As of 2019, "only seven countries are close to limiting emissions at the level necessary to keep warming to 2 degrees Celsius: India, the Philippines, Gambia, Morocco, Ethiopia, Costa Rica, and Bhutan." And only two, Morocco and the Gambia, were meeting the target of keeping warming to 1.5 degrees Celsius.[53] Notably absent on this list are the world's most powerful countries and major carbon emitters.

Why has it been so hard to implement effective policy in the United States?

One of the biggest tangible effects of the climate activism of the 1980s was to set off the fossil fuel industry reaction. Exxon and the American Petroleum Institute began to outline their counterattack at the end of the decade, beginning what would become a major lobbying and public relations campaign to cast doubt on the science and emphasize the economic risks of any significant action. They were not alone, because virtually the entire economy of the United States depended on fossil fuels. The National Association of Manufacturers, the US Chamber of Commerce, and associations for the coal, energy, and automobile industries soon jumped on board. By the mid-1990s, these organizations were spending tens of millions of dollars to counter science and every political attempt to take action.

The US infatuation with overconsumption predates and transcends the fossil fuel campaigns. In *The United States of Excess: Gluttony and the Dark Side of American Exceptionalism*, Robert Paarlberg explained some of the particularities that have made the fundamental political change we need particularly difficult.[54] The United States is a global outlier in its excessive consumption of both food and fossil fuels, leading to unusually high rates of obesity and per capita carbon emissions. Other industrialized countries have been able to enact policies to regulate the fast-food/junk-food industries and fossil fuel industries, allowing them to achieve significantly lower rates of both obesity and emissions. Why, Paarlberg asks, has the United States failed to enact comparable policies?

He focuses on three areas that distinguish the United States from other wealthy countries in its reluctance to regulate consumption: its material and demographic endowments, its unusual political structures, and its culture of individualism.[55] In that it alone escaped the massive material destruction of two world wars, the United States enjoyed an additional twentieth-century advantage over other industrialized countries (e.g., European countries and Japan).

In terms of material endowments, Paarlberg points to the country's enormous fossil fuel reserves and its low population density. Low density facilitates sprawl and high use of energy in transportation and shipping; large reserves give the fossil fuel industry outsized political influence. US political structures, including a weak federal government, bicameral legislatures, and a strong judiciary create multiple veto points, making it difficult to enact legislation that challenges special interests. Finally, the cultures of other industrialized countries (European countries and Japan) tend to value social rights more strongly, whereas in the US, the balance tilts towards individual rights. The result is that corporations in the United States enjoy more political power and less taxation and regulation than in Europe and Japan—and that we consume more junk food and fossil fuels than our counterparts there.

But while the United States has been unusually recalcitrant on economic and climate issues, no industrialized country has come close to rising to the challenges that face our world. Almost 1.6 trillion metric tons of carbon have spewed into the atmosphere since the Industrial Revolution began around 1750. The first 400 billion were released between 1750 and 1967; the second between 1968 and 1990; the third between 1991 and 2006; and the fourth between 2007 and 2018.[56] It's difficult to conclude we are making any progress at all.

CONCLUSION: REASONS FOR OPTIMISM

Despite this rather dreary history and the domestic and global structural obstacles, there are reasons for optimism. We may have understood the science of climate change in 1979, but today, people are recognizing the urgency and are showing willingness to act. One thing we didn't know in 1979—that we know now—was just how fast carbon emissions would increase if nothing was done to change our global trajectory. We are also directly witnessing the impacts of climate change, and we know how close we are to tipping points that may accelerate the process catastrophically and irreversibly.[57]

We have also made progress in articulating concrete policy measures that would go far beyond the timid and unenforced international agreements that have thus far failed to make a dent in global emissions. The framework of the Green New Deal gives us some radical, concrete, aspirational, yet achievable goals to fight for. A degrowth approach can liberate us to imagine a different, low-carbon, more just, and better world.

It is especially youth worldwide who are proving willing to acknowledge and mobilize for the kind of deep structural change that we need. Like the inhabitants of Third World countries and especially small island nations, they know that they are, and will be, the victims of others' policies and insouciance.

CONCLUSION

Fossil fuels are deeply woven into our industrial society and global economy. They played a central role in creating the material comforts many of us take for granted, as well as the harrowing inequalities that characterize our world. And their continued use threatens human life as we know it on this planet.

Climate justice means recognizing climate change as a moral, political, and economic issue that requires fundamentally reorganizing our global society and economy, not just a question of tweaking incentives and adding technologies. But how can we do it?

Who will lead?

It makes sense for governments to take the lead in the kind of global changes we need to combat climate change. Governments are the only institutions with a global presence and power, institutional mechanisms that they can work through, and at least in theory, accountability to their populations. Corporations may have global presence, power, and institutional mechanisms, but they are accountable only to their shareholders.

This lack of accountability doesn't mean that we should let corporations off the hook—there are lots of ways that the public can try to force corporations to change their ways, from fighting for government regulation to engaging in protest, boycott, and direct action. But we can't expect corporations to take the lead.

But governments and the international institutions they have created aren't taking the lead either. Many factors are hindering governments from taking the kind of steps we need to take. The most powerful of the world's governments represent the countries, and generally, the corporations and people within those countries, that are the most guilty—and they have no desire to punish themselves. Government accountability to their populations often takes the form of a narrow nationalism that they believe will boost their own stability and popularity rather than any larger interest in the future of humanity. Governments are also beholden to and dependent on corporations with little interest in the necessary change.

Both governments and corporations are further weighed down by bureaucratic inertia and reluctance to challenge the status quo. But challenging the status quo is exactly what is necessary right now. Understanding how the economic system that has brought us climate change is imbedded in our social, racial, and economic institutions will be key to challenging these institutions.

In order to envision and demand this kind of change, popular movements need to think about how the different actors are directly and indirectly complicit in the fossil fuel economy, what are the multiple points from which we can fight for change, and what kind of world we want to build to replace the fossil-dependent and unequal reality of today. The world's most powerful governments and corporations will not act if they are not pressured from below.

The IPCC urges 1.5

The 2015 Paris Agreement urged that the world's countries commit to policies to limit warming to well below 2 degrees Celsius, with the aim of 1.5 degrees Celsius. In a 2018 report, the IPCC made it clear the difference between the two would be vast. Even with a rise of 1.5 degrees Celsius, the impacts in terms of ocean temperature and marine life, extreme weather, drought, rising sea levels and floods, species loss and extinction, drought, and desertification, to name just a few, would be devastating. At 2 degrees Celsius, they would be much worse. The report stressed the urgency of the 1.5 degrees Celsius target, citing the ever-increasing "climate-related risks to health, livelihoods, food security, water supply, human security, and economic growth."[1]

Instead, the world's remaining carbon budget is rapidly being used up as emissions continued their upward trajectory. The IPCC concluded that to change course and achieve the 1.5-degrees Celsius goal "would require rapid and far-reaching transitions" in virtually every sector of society: "in energy, land, urban and infrastructure (including transport and buildings), and industrial systems. These systems transitions are unprecedented in terms of scale . . . and imply deep emissions reductions in all sectors, a wide portfolio of mitigation options and a significant upscaling of investments in those options."[2]

But the report kept its policy recommendations somewhat vague and avoided delving into the nature of these transformative changes or proposals about how to bring them about. That's not really the job of the scientists. But these political questions are exactly what we need to address.

One way to lower emissions: Pandemic shutdowns

As economies shut down worldwide during the COVID-19 pandemic, CO2 emissions plummeted. Factories closed, air travel halted, highways emptied. Emissions dropped at an unprecedented rate in the first four months of 2020. This was in the context of a steady upward trajectory over many decades, with a much smaller dip only during the 2008 recession. The 2020 drop actually put the world, for the first time, close to a course to meet the Paris goal—but only if the trajectory were to continue. 2020 emissions showed a 6.4 percent drop from 2019. But to attain the 1.5 degrees Celsius target, scientists estimate that global emissions would have to fall 7.6 percent every year between 2020 and 2030. Instead, emissions quickly rebounded as economies started to reopen.[3]

US emissions dropped the most: 13 percent, mostly from the decline in transportation and industry. Aviation emissions fell by 48 percent, and commuting ground to a halt. The United States accounted for a quarter of the global emissions decline: 650 million tons.[4]

The connection between emissions and the economy was never more obvious. But the emergency shutdown in the midst of a global pandemic brought the exact opposite of a just transition. Instead, the pain fell most heavily on the poor, while the very rich speculated and hoarded.

For many of the world's poor, already living day-to-day or week-to-week, the collapse was devastating. Some lost jobs, income, and subsistence. Others, "essential workers," faced the impossible choice of daily exposure to the virus to support their families. Those who suffered most from the shutdowns also suffered most from the virus.

"Pandemic profiteers"—some of the world's wealthiest people—did just fine during the shutdowns. During the pandemic year of 2020–21, the wealth of the US's six hundred-some billionaires grew even faster than its accelerated rate in previous decades. Some of the country's most profitable companies benefited from government largesse and new opportunities, and their shareholders made out handsomely.[5]

Under the worst of circumstances, we lowered emissions temporarily. The poor suffered the double attack of a pandemic and job loss while the rich were able to shield themselves from infection and economic collapse. But something else emerges from the statistics. The main factors behind falling emissions were drops in commuting, industrial production, and air travel. What if we prioritized restoring income and health, while acknowledging that declines in commuting, production, and air travel are actually a good thing? This would mean assimilating some of the key insights of the degrowth movement: what benefits the profits and consumption of the rich is very different from what will guarantee the rights of the poor.

Could we, with planning and redistribution, move to a slower and reconfigured economy in ways that don't cause the hardship, desperation, and hunger that the unplanned COVID-19 shutdowns entailed? Could we imagine a just transition to a different kind of economy that prioritizes sustainable redistribution to fulfill basic human needs?

Imagine, and act

The failure of several decades of top-down reforms, from the Kyoto Protocol of 1997 through the Paris Agreement of 2015, to make any serious dent in our ever-increasing global emissions should be evidence enough that our current institutions are not capable, on their own, of imagining or enacting the kind of changes we need.

As I've tried to show, from a technical standpoint there are many components to the climate emergency. Electricity generation, transportation, agriculture, housing, war, air travel, international trade, and the industrial-factory system all consume massive amounts of fossil fuel. The fossil fuel industry itself, and the industries and institutions that rely on it for their profits, collectively have a strong interest in maintaining the current system, with perhaps a few tweaks to enhance their public image.

The system relies on us buying its products and offering our implicit consent to business as usual. Most of us, with plenty of things to worry about in our personal lives, are tempted to leave the climate crisis to the experts—scientists, governments, and industry—to resolve. At most, we'll try to adjust our own consumption to buy greener.

But that's not going to work. Our institutions are too deeply imbedded in and dependent on the current institutional arrangements and economic system to even begin to consider the kind of major, structural, radical change that we need to have a habitable earth for our grandchildren.

Science and technology will necessarily play an important role in the transition we need, as will specific policies like regulations, caps, and changes in the tax structures and subsidies that currently actively promote GHG-emitting industries and activities. But I hope I've shown that without larger institutional, political, and social change, new technologies will only reproduce the very arrangements that got us into the current crisis. If we continue to privilege the profits and consumption of the world's elite, we will continue to abuse and exploit both humans and nature.

The past few years have seen a remarkable upsurge in vision and activism, and a synergy between the two. It's hard to mobilize to fight without a vision of the kind of social change you want, and when you have that vision, it's hard *not* to mobilize to bring it about.

The proposals put forth by the degrowth movement, the *buen vivir* movement, and the Green New Deal for Europe for an economic slowdown predicated on redistribution and social justice have begun to make that alternative vision much more concrete. Youth, and communities on the front lines, already experiencing the ravages of the warming climate, are bringing increasing popular and political urgency to the dire warnings of scientists.[6]

The problem may seem huge. But a global recognition of the urgency, magnitude, and interrelated nature of the problems is growing rapidly, pushed forward by the global peasant coalition Via Campesina, worldwide organizations demanding that we stop fossil fuels and that the rich countries contribute their "fair share," and the hundreds of thousands who have mobilized to protest the moribund international climate meetings that have failed thus far to reduce global emissions. There are plenty of avenues for getting involved. I hope this book will inspire readers to do so.

"There is no alternative" to the capitalist world economy, British prime minister Margaret Thatcher proclaimed triumphantly as the socialist world crumbled in the 1980s. But she was wrong, and today's climate activists are turning her phrase on its head: we have no alternative but to change course, radically and rapidly.

ACKNOWLEDGMENTS

Thanks to Steve Striffler, Noel Healy, Harry Chomsky, Michael Meeropol, and Alaitz Aritza for reading the manuscript and helping me clarify my ideas. Thanks to Justin Wolfe for organizing the Pandemic Social Distance Writing Group, which kept me going through multiple rounds of revisions. And thanks to Gayatri Patnaik, Amy Caldwell, and everyone at Beacon Press who helped shepherd this book into being.

NOTES

INTRODUCTION

1. "Confronting Carbon Inequality," Oxfam International, September 23, 2020, https://www.oxfam.org/en/research/confronting-carbon-inequality.

2. Stockholm Resilience Centre, "Planetary Boundaries," 2015, https://www.stockholmresilience.org/research/planetary-boundaries.html.

3. J. Rockström, W. Steffen, K. Noone, Å. Persson, F. S. Chapin III, E. Lambin, T. M. Lenton, M. Scheffer, C. Folke, H. Schellnhuber, B. Nykvist, C. A. De Wit, T. Hughes, S. van der Leeuw, H. Rodhe, S. Sörlin, P. K. Snyder, R. Costanza, U. Svedin, M. Falkenmark, L. Karlberg, R. W. Corell, V. J. Fabry, J. Hansen, B. Walker, D. Liverman, K. Richardson, P. Crutzen, and J. Foley, "Planetary Boundaries: Exploring the Safe Operating Space for Humanity," *Ecology and Society* 14, no. 2 (2009): 32, http://www.ecologyandsociety.org/vol14/iss2/art32.

4. Rob Wallace, *Big Farms Make Big Flu: Dispatches on Agriculture, Infectious Disease, and the Nature of Science* (New York: Monthly Review Press, 2016); R. Gibb, D. W. Redding, K. Q. Chin, et al., "Zoonotic Host Diversity Increases in Human-Dominated Ecosystems," *Nature* (2020).

CHAPTER 1: TECHNICAL QUESTIONS

1. International Society for Ecological Economics, http://www.isecoeco.org.

2. System Change, Not Climate Change, https://systemchangenotclimatechange.org.

3. US Environmental Protection Agency, "Overview of Greenhouse Gases," https://www.epa.gov/ghgemissions/overview-greenhouse-gases.

4. NASA Earth Observatory, "Methane Matters," Mar. 8, 2016, https://earthobservatory.nasa.gov/features/MethaneMatters; Environmental Defense Fund, "Methane: The Other Important Greenhouse Gas," https://www.edf.org/climate/methane-other-important-greenhouse-gas.

5. US Environmental Protection Agency, "Overview of Greenhouse Gases: Methane Emissions," https://www.epa.gov/ghgemissions/overview-greenhouse-gases#methane; Anthony J. Marchese and Dan Zimmerle, "Why Methane Emissions Matter to Climate Change," *The Conversation*, Sept. 4, 2019, https://theconversation.com/why-methane-emissions-matter-to-climate-change-5-questions-answered-122684; Jonathan Watts, "Oil and Gas Firms 'Have Had Far Worse Climate Impact Than Thought,'" *Guardian*, Feb. 19, 2020; Robert W. Howarth, "Ideas and Perspectives: Is Shale Gas a Major Driver of Recent Increase in Global

Atmospheric Methane?" *Biogeosciences* 16 (2019): 3033–46, https://www.biogeo sciences.net/16/3033/2019.

6. Hannah Ritchie and Max Roser, "Greenhouse Gas Emissions," Our World in Data, 2021, https://ourworldindata.org/greenhouse-gas-emissions. A metric ton is the equivalent of 1,000 kilograms, or a little over 2,200 pounds, slightly more than a US ton, which is 2,000 pounds. International reporting of climate data uses the metric system, and figures in this book generally follow that convention.

7. US Environmental Protection Agency, "Sources of Greenhouse Gas Emissions," https://www.epa.gov/ghgemissions/sources-greenhouse-gas-emissions; US Energy Information Administration, "Energy Use in Industry," https://www.eia .gov/energyexplained/use-of-energy/industry.php; US Environmental Protection Agency, "Sources of Greenhouse Gas Emissions: Industry," https://www.epa.gov /ghgemissions/sources-greenhouse-gas-emissions.

8. See Tom DiChristopher, "New Construction Natural Gas Ban Trend Appears on East Coast," *S&P Global*, Nov. 18, 2019.

9. US Energy Information Administration, "What Is U.S. Electricity Generation by Energy Source?" https://www.eia.gov/tools/faqs/faq.php?id=427&t=3.

10. US Energy Information Administration, "Electric Power Monthly: Net Generation by Energy Source, 2009-November 2019," Jan. 20, 2020, https://www .eia.gov/electricity/monthly/epm_table_grapher.php?t=epmt_1_01.

11. Trefis Team, "Chevron Picks a Fine Time to Leave the Coal Mining Biz," *Forbes*, Mar. 8, 2011; Maggie Borman, "ExxonMobil Sells Monterey Mine to the Cline Group," *Telegraph*, Jan. 7, 2009; Staff Reporter, "Exxon to Exit Australian Coal," *Australia's Mining Monthly*, July 24, 2000, https://www.miningmonthly .com/markets/international-coal-news/1302239/exxon-exit-australian-coal.

12. US Environmental Protection Agency, "Sources of Greenhouse Gas Emissions: Transportation," https://www.epa.gov/ghgemissions/sources-greenhouse -gas-emissions; US EPA, "Fast Facts on Transportation Greenhouse Gas Emissions," https://www.epa.gov/greenvehicles/fast-facts-transportation-greenhouse -gas-emissions.

13. US EPA, "Sources of Greenhouse Gas Emissions: Transportation."

14. Conor Brondson, "Electric Cars Won't Solve Climate Change," *Planetizen*, Mar. 9, 2021, https://www.planetizen.com/blogs/112490-electric-cars-wont-solve -climate-change.

15. See US EIA, "Energy Use in Industry," https://www.eia.gov/energyexplained /use-of-energy/industry.php#:ffl:text=The%20United%20States%20is%20a,of%20 total%20U.S.%20energy%20consumption.

16. Matthew T. Huber, "Hidden Abodes: Industrializing Political Ecology," *Annals of the American Association of Geographers* 107, no. 1 (2017): 151–66, 159, 162, 163.

17. Lucy Rodgers, "Climate Change: The Major Emitter You May Not Know About," *BBC News*, Dec. 17, 2018, https://www.bbc.com/news/science-environment -46455844.

18. Zhu Liu, "National Carbon Emissions from the Industry Process: Production of Glass, Soda Ash, Ammonia, Calcium Carbide and Alumina," *Applied Energy* 166 (Mar. 2016): 239–44.

19. See P. R. Shukla, J. Skea, E. Calvo Buendia, V. Masson-Delmotte, H.-O.

Pörtner, D. C. Roberts, P. Zhai, R. Slade, S. Connors, R. van Diemen, M. Ferrat, E. Haughey, S. Luz, S. Neogi, M. Pathak, J. Petzold, J. Portugal Pereira, P. Vyas, E. Huntley, K. Kissick, M. Belkacemi, J. Malley, eds., "Summary for Policymakers," in *Climate Change and Land: An IPCC Special Report on Climate Change, Desertification, Land Degradation, Sustainable Land Management, Food Security, and Greenhouse Gas Fluxes in Terrestrial Ecosystems* (New York: IPCC, 2019), https://www.ipcc.ch/srccl, 8.

20. Aleksandra Arcipowska, Emily Mangan, You Lyu, and Richard Waite, "5 Questions About Agricultural Emissions, Answered," World Resources Institute, July 29, 2019, https://www.wri.org/blog/2019/07/5-questions-about-agricultural-emissions-answered; Andrea Thompson, "Here's How Much Food Contributes to Climate Change," *Scientific American*, Sept. 13, 2021.

21. Nick Buxton, "The Elephant in Paris: The Military and Greenhouse Gas Emissions," TransNational Institute, Nov. 25, 2015, https://www.tni.org/es/node/22587; Arthur Nelsen, "Why the U.S. Military Is Losing Its Carbon-Emissions Exemption," *Atlantic*, Dec. 15, 2015; Oliver Belcher, Patrick Bigger, Benjamin Neimark, and Cara Kennelly, "Hidden Carbon Costs of the 'Everywhere War': Logistics, Geopolitical Ecology, and the Carbon Boot-Print of the US Military," *Transactions of the Institute of British Geographers* (June 2019): 1–16.

22. US Department of Energy, "Comprehensive Annual Energy Data and Sustainability Performance," FY 2018, Department of Defense, https://ctsedwweb.ee.doe.gov/Annual/Report/ComprehensiveGreenhouseGasGHGInventoriesByAgencyAndFiscalYear.aspx; Nikki Reisch and Steven Kretzmann, "A Climate of War: The War in Iraq and Global Warming," Oil Change International, Mar. 2008, http://priceofoil.org/content/uploads/2008/03/A%20Climate%20of%20War%20FINAL%20(March%2017%202008).pdf.

23. Neta C. Crawford, "Pentagon Fuel Use, Climate Change, and the Costs of War," Watson Institute, Brown University, June 12, 2019, https://watson.brown.edu/costsofwar/files/cow/imce/papers/2019/Pentagon%20Fuel%20Use,%20Climate%20Change%20and%20the%20Costs%20of%20War%20Final.pdf; Belcher et al., "Hidden Carbon Costs of the 'Everywhere War.'"

24. Crawford, "Pentagon Fuel Use, Climate Change, and the Costs of War."

25. Benjamin Hulac, "Pollution from Planes and Ships Left Out of Paris Agreement," *Scientific American*, Dec. 14, 2014.

26. Brandon Graver, Kevin Zhang, and Dan Rutherford, "CO2 Emissions from Commercial Aviation, 2018," International Council on Clean Transportation Working Paper 2019-16, https://theicct.org/sites/default/files/publications/ICCT_CO2-commercl-aviation-2018_20190918.pdf.

27. Graver, Zhang, and Rutherford, "CO2 Emissions"; Niko Kommenda, "How Your Flight Emits as Much CO2 as Many People Do in a Year," *Guardian*, July 19, 2019; Niall McCarthy, "The Worst Offenders for Air Travel Emissions," Statista.com, Oct. 22, 2019, https://www.statista.com/chart/19714/global-share-of-commercial-air-travel-emissions; Hiroko Tabuchi, "Should Flying Make You Feel Guilty?" *New York Times*, Oct. 22, 2019.

28. See Christopher Muther, "As 'Flight Shame' Movement Grows, More Airlines and Travelers Seek to Offset Carbon Footprint," *Boston Globe*, Jan. 25, 2020.

29. H. Res. 109, "Recognizing the Duty of the Federal Government to Create a Green New Deal," Feb. 7, 2019, https://www.congress.gov/116/bills/hres109/BILLS-116hres109ih.pdf; *White House Briefing Room Fact Sheet*, Jan. 27, 2021, https://www.whitehouse.gov/briefing-room/statements-releases/2021/01/27/fact-sheet-president-biden-takes-executive-actions-to-tackle-the-climate-crisis-at-home-and-abroad-create-jobs-and-restore-scientific-integrity-across-federal-government.

30. Hannah Beech, "Damming the Lower Mekong, Devastating the Ways and Means of Life," *New York Times*, Feb. 15, 2020; Bridget R. Deemer, John A. Harrison, Siyue Li, Jake J. Beaulieu, Tonya DelSontro, Nathan Barros, José F. Bezerra-Neto, Stephen M. Powers, Marco A. dos Santos, and J. Arie Vonk, "Greenhouse Gas Emissions from Reservoir Water Surfaces: A New Global Synthesis," *BioScience* 66, no. 11 (Nov. 1, 2016): 949–64, https://academic.oup.com/bioscience/article/66/11/949/2754271; Konrad Yakabuski, "The Dirty News About 'Clean' Hydropower Projects," *Globe and Mail*, Oct. 10, 2016.

31. Matt Hongoltz-Hetling, "Indigenous Activists Fight Expansion of Canadian Hydropower," *VTDigger*, Dec. 8, 2019, https://vtdigger.org/2019/12/08/indigenous-activists-fight-expansion-of-canadian-hydropower.

32. New England Clean Energy Connect, "About the Project," https://www.necleanenergyconnect.org/project-overview; Natural Resources Council of Maine, "CMP Transmission Line Proposal: A Bad Deal for Maine," https://www.nrcm.org/programs/climate/proposed-cmp-transmission-line-bad-deal-maine.

33. Morgan Lowrie, "New York Officials Tour Quebec Indigenous Communities Ahead of Possible Hydro Deal," *Canada's National Observer*, July 31, 2019, https://www.nationalobserver.com/2019/07/31/news/new-york-officials-tour-quebec-indigenous-communities-ahead-possible-hydro-deal.

34. National Renewable Energy Laboratory, "Life Cycle Assessment Harmonization," https://www.nrel.gov/analysis/life-cycle-assessment.html.

35. Angela Chen, "More Solar Panels Mean More Waste and There's No Easy Solution," *The Verge*, Oct. 25, 2018, https://www.theverge.com/2018/10/25/18018820/solar-panel-waste-chemicals-energy-environment-recycling; Michael Shellenberger, "If Solar Panels Are So Clean, Why Do They Produce So Much Toxic Waste?" *Forbes*, May 23, 2018.

36. National Renewable Energy Laboratory, "Life Cycle Assessment Harmonization."

37. See Lee M. Miller and David W. Keith, "Observation-Based Solar and Wind Power Capacity Factors and Power Densities, *Environmental Research Letters* 13, no. 10 (Oct. 4, 2018), https://iopscience.iop.org/article/10.1088/1748-9326/aae102; Mark Z. Jacobson, "Response to Miller and Keith," Oct. 3, 2018, unpublished ms., available at https://web.stanford.edu/group/efmh/jacobson/Articles/I/CombiningRenew/18-RespERL-MK.pdf.

38. Thea Riofrancos, "What Green Costs," *Logic*, Dec. 7, 2019, https://logicmag.io/nature/what-green-costs.

39. Naveena Sadasivam, "Report: Going 100% Renewable Power Means a Lot of Dirty Mining," *Grist*, Apr. 17, 2019, https://grist.org/article/report-going-100-renewable-power-means-a-lot-of-dirty-mining; Riofrancos, "What Green Costs";

UNCTAD, "Commodities at a Glance: Special Issue on Strategic Battery Raw Materials" (United Nations, 2020), 46, https://unctad.org/en/PublicationsLibrary/ditccom2019d5_en.pdf; Justine Calma, "The Electric Vehicle Industry Needs to Figure Out Its Battery Problem," *The Verge*, Nov. 6, 2019, https://www.theverge.com/2019/11/6/20951807/electric-vehicles-battery-recycling; Amit Katwala, "The Spiraling Environmental Cost of Our Lithium Battery Addiction," *Wired*, Aug. 5, 2018, https://www.wired.co.uk/article/lithium-batteries-environment-impact.

40. Heather Rogers, *Green Gone Wrong: Dispatches from the Front Lines of Eco-Capitalism* (London: Verso, 2013), 181.

41. For a full analysis, see US Environmental Protection Agency, *Biofuels and the Environment: The Second Triennial Report to Congress*, EPA/600/R-18/195 (Washington, DC: US EPA, 2018), https://cfpub.epa.gov/si/si_public_record_report.cfm?Lab=IO&dirEntryId=341491.

42. Raj Patel, foreword, in Timothy A. Wise, *Eating Tomorrow: Agribusiness, Family Farmers, and the Battle for the Future of Food* (New York: New Press, 2019).

43. Hannah Ritchie and Max Roser, "Access to Energy," Our World in Data, 2019, https://ourworldindata.org/energy-access.

44. James Dyke, Robert Watson, and Wolfgang Knorr, "Climate Scientists: Concept of Net Zero Is a Dangerous Trap," *The Conversation*, Apr. 22, 2021, https://theconversation.com/climate-scientists-concept-of-net-zero-is-a-dangerous-trap-157368.

45. Taran Volckhausen, "Forests Scramble to Absorb Carbon as Emissions Continue to Increase," *Mongabay*, Mar. 21, 2019, https://news.mongabay.com/2019/03/forests-scramble-to-absorb-carbon-as-emissions-continue-to-increase.

46. One hectare is approximately 2.5 acres.

47. UN Food and Agriculture Organization, "Global Forest Resource Assessment 2020: Key Findings," http://www.fao.org/documents/card/en/c/ca8753en.

48. Mikaela Weisse and Liz Goldman, "The World Lost a Belgium-Sized Area of Primary Rainforests Last Year," *Global Forest Watch*, Apr. 25, 2019, https://blog.globalforestwatch.org/data-and-research/world-lost-belgium-sized-area-of-primary-rainforests-last-year; Volckhausen, "Forests Scramble to Absorb Carbon."

49. Adam Aton, "Surprisingly, Tropical Forests Are Not a Carbon Sink," *Scientific American*, Sept. 29, 2017; David Gibbs, Nancy Harris, and Frances Seymour, "By the Numbers: The Value of Tropical Forests in the Climate Change Equation," World Resources Institute, Oct. 4, 2018; Volckhausen, "Forests Scramble to Absorb Carbon."

50. Georgina Gustin, "Planes Sampling Air Above the Amazon Find the Rainforest Is Releasing More Carbon Than It Stores," *Inside Climate News*, July 14, 2021, https://insideclimatenews.org/news/14072021/amazon-deforestation-climate-change-brazil-carbon-source.

51. Rachel Fritts, "Tropical Deforestation Now Emits More CO2 Than the EU," *Mongabay*, Oct. 18, 2018, https://news.mongabay.com/2018/10/tropical-deforestation-now-emits-more-co2-than-the-eu; UN Food and Agriculture Organization, *Forest Land Emissions and Removals: Global and Country Trends, 1990–2020*, FAOSTAT Analytical Brief Series No. 12 (Rome: FAO, 2020), http://www.fao.org/3/cb1578en/CB1578EN.pdf.

52. "Capturing Carbon's Potential: These Companies Are Turning CO2 into Profits," *State of the Planet* (blog), May 29, 2019, https://blogs.ei.columbia.edu/2019/05/29/co2-utilization-profits.

53. Akshat Rathi, "The EU Has Spent Nearly $500 Million on Technology to Fight Climate Change—with Little to Show for It," *Quartz*, Oct. 23, 2018, https://qz.com/1431655/the-eu-spent-e424-million-on-carbon-capture-with-little-to-show-for-it; Bert Metz, Ogunlade Davidson, Heleen de Coninck, Manuela Loos, and Leo Meyer, eds., *IPCC Special Report on Carbon Dioxide Capture and Storage* (Cambridge: Cambridge University Press, 2005), https://www.ipcc.ch/report/carbon-dioxide-capture-and-storage.

54. Metz et al., "Summary for Policymakers," *IPCC Special Report on Carbon Dioxide Capture and Storage*, 3; V. Masson-Delmotte, P. Zhai, H.-O. Pörtner, D. Roberts, J. Skea, P.R. Shukla, A. Pirani, W. Moufouma-Okia, C. Péan, R. Pidcock, S. Connors, J.B.R. Matthews, Y. Chen, X. Zhou, M.I. Gomis, E. Lonnoy, T. Maycock, M. Tignor, and T. Waterfield, eds., "Summary for Policymakers," in *Global Warming of 1.5°C: An IPCC Special Report on the Impacts of Global Warming of 1.5°C Above Pre-Industrial Levels and Related Global Greenhouse Gas Emission Pathways, in the Context of Strengthening the Global Response to the Threat of Climate Change, Sustainable Development, and Efforts to Eradicate Poverty* (Geneva: IPCC, 2018), 14, 17, https://www.ipcc.ch/site/assets/uploads/sites/2/2019/05/SR15_SPM_version_report_LR.pdf.

55. Rachel M. Cohen, "The Environmental Left Is Softening on Carbon Capture Technology. Maybe That's OK," *The Intercept*, Sept. 20, 2019, https://theintercept.com/2019/09/20/carbon-capture-technology-unions-labor.

56. See World Coal Organization, "Carbon Capture, Use and Storage," https://www.worldcoal.org/reducing-co2-emissions/carbon-capture-use-storage; ExxonMobil, "What If We Could Stop Carbon Dioxide Emissions from Power Plants?" May 5, 2016, https://energyfactor.exxonmobil.com/science-technology/stop-carbon-dioxide-emissions-power-plants. See AFL-CIO, Resolution 55, "Climate, Energy, and Union Jobs," Oct. 24, 2017, https://aflcio.org/resolutions/resolution-55-climate-change-energy-and-union-jobs; Patrick Falwell and Brad Crabtree, "Understanding the National Enhanced Oil Recovery Initiative," *The Cornerstone* (Winter 2014), https://www.c2es.org/document/understanding-the-national-enhanced-oil-recovery-initiative.

57. Carbon Capture Coalition, "About the Carbon Capture Coalition," https://carboncapturecoalition.org/about-us.

58. Friends of the Earth, "Groups Decry New Carbon Capture Subsidies," July 16, 2016, https://foe.org/news/2016-07-groups-decry-new-carbon-capture-subsidies.

59. James Temple, "The UN Climate Report Pins Hopes on Carbon Removal Technologies That Barely Exist," *MIT Technology Review*, August 9, 2021, https://www.technologyreview.com/2021/08/09/1031450/the-un-climate-report-pins-hopes-on-carbon-removal-technologies-that-barely-exist/; Jason Hickel et al., "Urgent Need for Post-Growth Climate Mitigation Scenarios," *Nature Energy* 6, no. 8 (August 2021): 766–68, https://doi.org/10.1038/s41560-021-00884-9.

60. US Environmental Protection Agency, "Sources of Greenhouse Gas Emissions: Electricity Sector Emissions," https://www.epa.gov/ghgemissions/sources-greenhouse-gas-emissions.

61. Brian Barth, "Is LEED Tough Enough for the Climate Change Era?" *CityLab,* June 5, 2018, https://www.citylab.com/environment/2018/06/is-leed-tough-enough-for-the-climate-change-era/559478.

62. Megan Darby, "'Green' Bond to Fund Multi-Storey Car Park," *Climate Home News,* Jan. 13, 2015, https://www.climatechangenews.com/2015/01/13/green-bond-to-fund-multi-storey-car-park; Mark Cherney, "'Green Bonds' for a Parking Garage?" *Wall Street Journal,* Mar. 12, 2015; Parksmart, "Salem State North Campus Transportation Center," https://parksmart.gbci.org/salem-state-north-campus-transportation-center, and USGBC, "Parksmart Enables Our Clients to Meet Their Sustainability Goals," Aug. 2, 2016, https://www.usgbc.org/articles/parksmart-enables-our-clients-meet-their-sustainability-goals.

63. Thomas Frank, with Christopher Schnaars and Hannah Morgan, "In U.S. Building Industry, Is It Too Easy to Be Green?" *USA Today,* Oct. 24, 2012 (updated June 13, 2013); Mireya Navarro, "Some Buildings Not Living Up to Green Label," *New York Times,* Aug. 30, 2009; Sam Roudman, "Bank of America's Toxic Tower," *New Republic,* July 29, 2013.

64. Frank, "Is It Too Easy to Be Green?"

CHAPTER 2: POLICY QUESTIONS

1. UN, Climate Change, Process and Meetings, The Kyoto Protocol, "Emissions Trading," https://unfccc.int/process/the-kyoto-protocol/mechanisms/emissions-trading.

2. S.Res. 98, 105th Congress (1997–1998), "A Resolution Expressing the Sense of the Senate Regarding the Conditions for the United States Becoming a Signatory to Any International Agreement on Greenhouse Gas Emissions Under the UN Framework Convention on Climate Change." https://www.congress.gov/bill/105th-congress/senate-resolution/98.

3. Anonymous, "A Greener Bush," *Economist,* Feb. 15, 2003.

4. Michael Le Page, "Was Kyoto Climate Deal a Success?" *New Scientist,* June 14, 2016, https://www.newscientist.com/article/2093579-was-kyoto-climate-deal-a-success-figures-reveal-mixed-results.

5. Climate Action Tracker: Countries, https://climateactiontracker.org/countries; International Energy Agency, "Global CO2 Emissions in 2019," https://www.iea.org/articles/global-co2-emissions-in-2019; IEA, "After Steep Drop in Early 2020, Global Carbon Dioxide Emissions Have Rebounded Strongly," Mar. 2, 2021, https://www.iea.org/news/after-steep-drop-in-early-2020-global-carbon-dioxide-emissions-have-rebounded-strongly.

6. Kieran Mulvaney, "Climate Change Report Card," *National Geographic,* Sept. 19, 2019.

7. Petteri Taalas and Joyce Msuya, foreword, in Masson-Delmotte et al., *Global Warming of 1.5°C,* https://www.ipcc.ch/sr15/about/foreword.

8. Greg Muttitt and Sivan Kartha, "Equity, Climate Justice, and Fossil Fuel Extraction: Principles for a Managed Phase Out," *Climate Policy* (2020).

9. European Commission, "EU Emissions Trading System," https://ec.europa.eu/clima/policies/ets_en.

10. European Commission, "EU Emissions Trading System."

11. OECD, "Few Countries Are Pricing Carbon High Enough to Meet Carbon Targets," Sept. 18, 2018, https://www.oecd.org/tax/few-countries-are-pricing-carbon-high-enough-to-meet-climate-targets.htm.

12. Acadia Center, "The Regional Greenhouse Gas Initiative: Ten Years in Review," Acadia Center, 2019, https://acadiacenter.org/wp-content/uploads/2019/09/Acadia-Center_RGGI_10-Years-in-Review_2019-09-17.pdf; Jan Ellen Spiegel, "Power Plant Emissions Down 47% Under the Regional Greenhouse Gas Initiative," Yale Climate Connections, Jan. 16, 2020, https://www.yaleclimateconnections.org/2020/01/power-plant-emissions-down-47-percent-under-the-regional-greenhouse-gas-initiative.

13. David Roberts, "The Northeast's Carbon Trading System Works Quite Well. It Just Doesn't Reduce Much Carbon," *Vox*, Feb. 28, 2017, https://www.vox.com/science-and-health/2017/2/28/14741384/rggi-explained.

14. David Roberts, "California's Cap-and-Trade System May Be Too Weak to Do Its Job," *Vox*, Dec. 13, 2018, https://www.vox.com/energy-and-environment/2018/12/12/18090844/california-climate-cap-and-trade-jerry-brown.

15. Roberts, "California's Cap-and-Trade System May Be Too Weak to Do Its Job"; Lisa Song, "Cap and Trade Is Supposed to Solve Climate Change, but Oil and Gas Industry Emissions Are Up," *ProPublica*, Nov. 15, 2019, https://www.propublica.org/article/cap-and-trade-is-supposed-to-solve-climate-change-but-oil-and-gas-company-emissions-are-up; Brad Plumer, "Just How Far Can California Possibly Go on Climate?" *New York Times*, July 26, 2017.

16. World Bank Group, "State and Trends of Carbon Pricing, 2019," http://documents.worldbank.org/curated/en/191801559846379845/pdf/State-and-Trends-of-Carbon-Pricing-2019.pdf.

17. "Encyclical Letter Laudato Si' of the Holy Father Francis on Care for Our Common Home," May 24, 2015, http://w2.vatican.va/content/francesco/en/encyclicals/documents/papa-francesco_20150524_enciclica-laudato-si.html.

18. Climate and Prosperity, "Cap and Dividend," http://climateandprosperity.org; Cap Global Carbon, "Cap Global Carbon," https://capglobalcarbon.org; Peter Barnes, "Cap and Dividend, Not Trade: Making Polluters Pay," *Scientific American*, Dec. 1, 2008.

19. Brad Plumer and Nadja Popovich, "These Countries Have Prices on Carbon. Are They Working?" *New York Times*, Apr. 2, 2019.

20. John Larsen, Shashank Mohan, Peter Marsters, and Whitney Herndon, "Energy and Environmental Implications of a Carbon Tax in the United States," Rhodium Group, July 17, 2018, https://energypolicy.columbia.edu/research/report/energy-and-environmental-implications-carbon-tax-united-states.

21. Center for Biological Diversity, "Carbon Dividend Bill in House Would Gut Clean Air Act Authority to Stop Climate Change," Jan. 25, 2019, https://www.biologicaldiversity.org/news/press_releases/2019/greenhouse-gas-emissions-01-25-2019.php.

22. In the first three years of the Trump administration, coal-fired plants producing 32 GW of electricity were closed. See Benjamin Storrow, "Global Emissions Were Flat in 2019—But Don't Cheer Yet," *Scientific American*, Feb. 12, 2020. Coal-fired power generation fell by 18 percent in 2019—the largest decline ever. See

Trevor Houser and Hannah Pitt, "Preliminary U.S. Emissions Estimates for 2019," Rhodium Group, Jan. 7, 2020, https://www.eenews.net/assets/2020/01/07/document_cw_01.pdf.

23. Miranda Green and Alex Gangitano, "Oil Companies Join Blitz for Carbon Tax," *The Hill*, May 22, 2019, https://thehill.com/policy/energy-environment/445100-oil-companies-join-blitz-for-carbon-tax; Marianne Lavelle, "Carbon Tax Plans: How They Compare and Why Oil Giants are Supporting One of Them," *Inside Climate News*, Mar. 7, 2019, https://insideclimatenews.org/news/07032019/carbon-tax-proposals-compare-baker-shultz-exxon-conocophillips-ccl-congress; Jim Walsh, "The Oil Industry's Climate Tax Dream Is a Climate Nightmare," *Food and Water Watch*, Oct. 7, 2019, https://www.foodandwaterwatch.org/news/oil-industrys-carbon-tax-dream-climate-nightmare.

24. Progressive organizations like the Green New Deal for Europe emphasize the redistributive nature of a fee-and-dividend system that essentially takes more from the rich (wealthy individuals and large corporations that are higher consumers) and gives more (proportionately, or absolutely, depending on its formulation) to the poor. Green New Deal for Europe, *A Blueprint for Europe's Just Transition*, 2nd ed., Dec. 2019, https://report.gndforeurope.com

25. Climate Leadership Council, https://clcouncil.org.

26. Jennifer A. Dlouhy and Ari Natter, "Oil Companies Join Corporate Lobbying Push for U.S. Carbon Tax," *Bloomberg*, May 20, 2019.

27. Kim Willsher, "Macron Scraps Fuel Tax Rise in Face of Gilets Jaunes Protests," *Guardian*, Dec. 5, 2018; Feargus O'Sullivan, "Why Drivers are Leading a Protest Movement Across France," *CityLab*, Nov. 19, 2018, https://www.citylab.com/transportation/2018/11/french-protests-gilets-jaunes-emmanuel-macron-gas-diesel-tax/576196.

28. UN Food and Agriculture Organization, "Forest," http://www.fao.org/3/Y4171E/Y4171E10.htm; Barbara Haya, "Policy Brief: The California Air Resources Board's U.S. Forest Offset Protocol Underestimates Leakage," Goldman School of Public Policy, University of California, Berkeley, May 7, 2019, https://gspp.berkeley.edu/assets/uploads/research/pdf/Policy_Brief-US_Forest_Projects-Leakage-Haya_4.pdf.

29. John Vidal, "Rich Countries to Pay Energy Giants to Build New Coal-Fired Power Plants," *Guardian*, July 14, 2010.

30. Kaisa Amaral, "What Does Flight Shame Have to Do with Global Carbon Markets?" *Climate Market Watch*, Nov. 28, 2019.

31. Indigenous Environmental Network, "UN Promoting Potentially Genocidal Policy at World Climate Summit," Dec. 8, 2015, https://www.ienearth.org/un-promoting-potentially-genocidal-policy-at-world-climate-summit; Julie Velásquez-Runk, "Creating Wild Darién: Centuries of Darién's Imaginative Geography and Its Lasting Effects," *Journal of Latin American Geography* 14, no. 3 (Oct. 2015): 127–56, 146.

32. Jason G. Goldman, "Ecuador Has Begun Drilling for Oil in the World's Richest Rainforest," *Vox*, Jan. 14, 2017, https://www.vox.com/energy-and-environment/2017/1/14/14265958/ecuador-drilling-oil-rainforest; Kimberly Brown, "Heart of Ecuador's Yasuní, Home to Uncontacted Tribes, Opens for Oil Drilling,"

Mongabay, July 5, 2019, https://news.mongabay.com/2019/07/heart-of-ecuadors-yasuni-home-to-uncontacted-tribes-opens-for-oil-drilling.

33. Shelagh Whitley, Han Chen, Alex Doukas, Ipek Gençsü, Ivetta Gerasimchuk, Yanick Touchette, and Leah Worrall, "Fossil Fuel Subsidy Scorecard: Methodology Note," Overseas Development Institute (ODI), with Natural Resources Defense Council, International Institute for Sustainable Development (IISD), and Oil Change International, June 2018, https://www.odi.org/sites/odi.org.uk/files/resource-documents/12218.pdf; Jude Clemente, "President Obama's Support for America's Shale Oil and Natural Gas," *Forbes*, Dec. 31, 2019.

34. Clayton Coleman and Emma Dietz, "Fact Sheet: Fossil Fuel Subsidies: A Closer Look at Tax Breaks and Societal Costs," Environmental and Energy Study Institute, July 29, 2019, https://www.eesi.org/papers/view/fact-sheet-fossil-fuel-subsidies-a-closer-look-at-tax-breaks-and-societal-costs; Sonali Prasad, Jason Burke, Michael Slezak, and Oliver Milman, "Obama's Dirty Secret: The Fossil Fuel Projects the US Littered Around the World," *Guardian*, Dec. 1, 2016.

35. Peter Erickson, Harro van Asselt, Doug Koplow, Michael Lazarus, Peter Newell, Naomi Oreskes, and Geoffrey Supran, "Why Fossil Fuel Subsidies Matter," *Nature* 578, no. 7793 (Feb. 5, 2020): E1–E4, E2.

36. David Coady, Ian Parry, Nghia-Piotr Le, and Baoping Shang, "Global Fossil Fuel Subsidies Remain Large: An Update Based on Country-Level Estimates," International Monetary Fund (IMF) Working Paper, May 2019, https://www.imf.org/en/Publications/WP/Issues/2019/05/02/Global-Fossil-Fuel-Subsidies-Remain-Large-An-Update-Based-on-Country-Level-Estimates-46509.

37. Jon Greenberg, "Did President Obama Save the Auto Industry?" *PolitiFact*, Sept. 6, 2012, https://www.politifact.com/truth-o-meter/article/2012/sep/06/did-obama-save-us-automobile-industry; Julie Young, "Too Big to Fail," *Investopedia*, Apr. 30, 2019, https://www.investopedia.com/terms/t/too-big-to-fail.asp.

38. Brad Plumer, "Obama Says Fracking Can Be a Bridge to a Clean Energy Future. It's Not That Simple," *Washington Post*, Jan. 29, 2014.

39. Emily Atkin, "Big Oil Wants Your Love, and It's Using Obama to Get It," *Heated*, Oct. 29, 2019. See the video at American Petroleum Institute, "U.S. Energy Empowers American Progress, Not Partisanship," https://youtu.be/CqYyBb5TA94.

40. US Energy Information Administration, "U.S. Coal Consumption in 2018 Expected to Be Lowest in 39 Years," Dec. 4, 2018, https://www.eia.gov/todayinenergy/detail.php?id=37692; US Department of Energy, Office of Fossil Energy, "Shale Research and Development," https://www.energy.gov/fe/science-innovation/oil-gas-research/shale-gas-rd.

41. Holmes Lybrand, "Fact Check: Biden Falsely Claims He Never Opposed Fracking," CNN, Oct. 23, 2020, https://www.cnn.com/2020/10/23/politics/biden-fracking-fact-check/index.html.

42. Adam Aton, "Biden's Promise to Unions: 'I'm All for Natural Gas,'" *E&E News*, Mar. 4, 2021, https://www.eenews.net/stories/1063726593; Ken Silverstein, "Natural Gas Is Core to the New Energy Economy," *Forbes*, Sept. 9, 2020; Associated Press, "Oil Companies Lock in Drilling, Challenging Biden on Climate Change," *Los Angeles Times*, Jan. 11, 2021; Branko Marcetic, "Joe Biden Is Almost as

Pro-Drilling as Trump," *Jacobin*, June 3, 2021, https://www.jacobinmag.com/2021/06/joe-biden-climate-policy-drilling-trump.

43. Hirochi Tabuchi, "Methane Leak, Seen from Space, Proves to Be Far Larger Than Thought," *New York Times*, Dec. 17, 2019; Jonah M. Kessel and Hirochi Tabuchi, "It's a Vast, Invisible Climate Menace. We Made It Visible," *New York Times*, Dec. 17, 2019, https://www.nytimes.com/interactive/2019/12/12/climate/texas-methane-super-emitters.html; Robert W. Howarth, "Ideas and Perspectives: Is Shale Gas a Major Driver of Recent Increase in Global Atmospheric Methane?" *Biogeosciences* 16 (2019), 3033–3046, https://www.biogeosciences.net/16/3033/2019.

44. Marchese and Zimmerle, "Why Methane Emissions Matter." Howarth gives the 3.5 percent estimate. For even higher estimates, see Hiroko Tabuchi, "Oil and Gas May Be a Far Bigger Climate Threat Than We Knew," *New York Times*, Feb. 19, 2020.

45. US Energy Information Administration, "EIA Expects Energy-Related CO2 Emissions to Fall in 2019," July 5, 2019, https://www.eia.gov/todayinenergy/detail.php?id=40094; US Energy Information Administration, "U.S. Liquefied Natural Gas Export Capacity to More Than Double by the End of 2019," Dec. 2018, https://www.eia.gov/todayinenergy/detail.php?id=37732.

46. Nicholas Kusnetz, "Natural Gas Rush Drives a Global Rise in Fossil Fuel Emissions," *Inside Climate News*, Dec. 3, 2019, https://insideclimatenews.org/news/03122019/fossil-fuel-emissions-2019-natural-gas-bridge-oil-coal-climate-change; Michael Levy, "Climate Consequences of Natural Gas as a Bridge Fuel," *Climatic Change* 118 (2013): 609–23, https://link.springer.com/article/10.1007/s10584-012-0658-3?no-access=true#author-information; American Petroleum Institute, "Climate Action Framework," 2021, https://www.api.org/-/media/Files/EHS/climate-change/2021/api-climate-action-framework.pdf?la=en&hash=E6BB3FA3013B52153E10D3E66C52616E00411D20.

47. Naomi Oreskes and Erik M. Conway, *Merchants of Doubt: How a Handful of Scientists Obscured the Truth on Issues from Tobacco Smoke to Global Warming* (New York: Bloomsbury, 2010), 245; Geoffrey Suprans and Naomi Oreskes, "Assessing Exxon-Mobil's Climate Change Communications (1977–2014)," *Environmental Research Letters* 12, no. 8 (2017): 1–18.

48. Elaine Kamarck, "The Challenging Politics of Climate Change," Brookings Institution, Sept. 23, 2019; Matthew Ballew, Abel Gustafson, Parrish Bergquist, Matthew Goldberg, Seth Rosenthal, John Kotcher, Edward Maibach, and Anthony Leiserowitz, "Americans Underestimate How Many Others in the U.S. Think Global Warming Is Happening," Yale University and George Mason University. New Haven, CT: Yale Program on Climate Change Communication, July 2, 2019, https://climatecommunication.yale.edu/publications/americans-underestimate-how-many-others-in-the-u-s-think-global-warming-is-happening; Stephen Johnson, "75 Percent of Americans Now Believe Humans Fuel Climate Change," *Big Think*, Sept. 16, 2019, https://bigthink.com/politics-current-affairs/climate-change-poll-americans?rebelltitem=1#rebelltitem1.

49. Donald Gutstein, *The Big Stall: How Big Oil and Think Tanks Are Blocking Action on Climate Change in Canada* (Toronto: James Lorimer & Company, 2018), 10.

50. Julie Doyle, "Where Has All the Oil Gone? BP Branding and the Discursive Elimination of Climate Change Risk," in *Culture, Environment and Eco-Politics*, ed. Nick Heffernan and David Wragg (Newcastle upon Tyne, UK: Cambridge Scholars Press, 2011), 200–225, 201.

51. ExxonMobil, "2020 Energy and Carbon Summary," https://corporate.exxonmobil.com/-/media/Global/Files/energy-and-carbon-summary/Energy-and-carbon-summary.pdf.

52. Gutstein, *The Big Stall*, 168.

53. Mary Annaise Heglar, "I Work in the Environmental Movement. I Don't Care If You Recycle," *Vox*, June 4, 2019, https://www.vox.com/the-highlight/2019/5/28/18629833/climate-change-2019-green-new-deal.

54. British Petroleum, "Know Your Carbon Footprint," https://www.knowyourcarbonfootprint.com.

55. Heglar, "I Work in the Environmental Movement."

56. Heather Rogers, *Gone Today: The Hidden Life of Garbage* (New York: New Press, 2005), chap. 6m, 143, 144.

57. Rogers, *Gone Today*, 4.

58. Kate Yoder, "Love It or Hate It, Earth Day's Just Not What It Used to Be. What Happened?" *Grist*, Apr. 22, 2019, https://grist.org/article/love-it-or-hate-it-earth-days-just-not-what-it-used-to-be-what-happened.

59. William Safire, "Footprint," *New York Times Magazine*, Feb. 17, 2008.

60. US Chamber of Commerce, Global Energy Institute, "What If . . . Hydraulic Fracturing Was Banned?" 2020 edition, https://www.globalenergyinstitute.org/energy-accountability.

61. Andreas Malm, *Fossil Capital: The Rise of Steam Power and the Roots of Global Warming* (London: Verso, 2016).

62. Matthew T. Huber, *Lifeblood: Oil, Freedom, and the Forces of Capital* (Minneapolis: University of Minnesota Press, 2013), xiv–xv.

63. Huber, *Lifeblood*, 29, 59, 56.

64. John Schwartz, "Judge Dismisses Suit Against Oil Companies over Climate Change Costs," *New York Times*, June 25, 2018.

65. Michael Brune, "Pulling Down Our Monuments," Sierra Club, July 22, 2020, https://www.sierraclub.org/michael-brune/2020/07/john-muir-early-history-sierra-club.

66. Genevieve LeBaron, "Green NGOs Cannot Take Big Business Cash and Save Planet," *The Conversation*, Sept. 30, 2013, https://theconversation.com/green-ngos-cannot-take-big-business-cash-and-save-planet-18770; Extinction Rebellion, https://rebellion.global; Sunrise Movement, "We Are the Climate Revolution," https://www.sunrisemovement.org/?ms=SunriseMovement.

67. "The Fossil Fuel Non-Proliferation Treaty," https://fossilfueltreaty.org/home; "Build Back Fossil Free," https://buildbackfossilfree.org.

68. David Roberts, "Introducing Climate Hawks," *Grist*, Oct. 21, 2010, https://grist.org/article/2010-10-20-introducing-climate-hawks.

69. Naomi Klein, *This Changes Everything: Capitalism Versus the Climate* (New York: Simon and Schuster, 2014); Kate Aronoff, Alyssa Battistoni, Daniel Aldana

Cohen, and Thea Riofrancos, *A Planet to Win: Why We Need a Green New Deal* (London: Verso, 2020).

70. Green New Deal letter to Congress, Jan. 10, 2019, http://foe.org/wp-content/uploads/2019/01/Progressive-Climate-Leg-Sign-On-Letter-2.pdf.

71. H. Res. 109, "Recognizing the Duty of the Federal Government to Create a Green New Deal," Feb. 7, 2019, https://www.congress.gov/116/bills/hres109/BILLS-116hres109ih.pdf.

72. David Roberts, "Fox News Has United the Right Against the Green New Deal. The Left Remains Divided," *Vox*, Apr. 22, 2019, https://www.vox.com/energy-and-environment/2019/4/22/18510518/green-new-deal-fox-news-poll.

73. "Trump Mocks Democrats' Green New Deal," *Axios*, Feb. 9, 2019, https://www.axios.com/green-new-deal-trump-tweets-democrats-socialism-12777aeb-7c7d-4790-88e8-cde49b5255b3.html.

74. Thea Riofrancos, "Reflections on the Green New Deal," *Viewpoint Magazine*, May 16, 2019, https://www.viewpointmag.com/2019/05/16/plan-mood-battlefield-reflections-on-the-green-new-deal; Asad Rehman, "The 'Green New Deal' Supported by Ocasio-Cortez and Corbyn Is Just a New Form of Colonialism," *Independent*, May 4, 2019; Aronoff, Battistoni, Cohen, and Riofrancos, *A Planet to Win*.

75. "The Biden Plan for a Clean Energy Revolution and Environmental Justice," https://joebiden.com/climate-plan; David Roberts, "What Joe Biden Was Trying to Say About the Green New Deal," *Vox*, Oct. 7, 2020, https://www.vox.com/energy-and-environment/21498236/joe-biden-green-new-deal-debate.

76. Umair Irfan, "Five Things to Know About the New U.S. Climate Commitment," *Vox*, Apr. 22, 2021, https://www.vox.com/22397364/earth-day-us-climate-change-summit-biden-john-kerry-commitment-2030-zero-emissions; David Roberts, "America Is Making Climate Promises Again. Should Anyone Care?" *Vox*, Apr. 27, 2021, https://www.vox.com/22401917/biden-climate-plan-summit-republicans-congress-midterm-elections.

77. Tom Athanasiou, "Over 50,000 People and 195 Global Groups Demand Biden Commit the US to Do Its 'Fair Share' on Climate," US Climate Fair Share, Feb. 21, 2021, https://usfairshare.org/blog/over-50000-people-195-global-groups-demand-biden-commit-the-u-s-to-do-its-fair-share-on-climate.

78. Kate Aronoff, "The Big Difference Between a Green New Deal and Biden's Climate Agenda," *New Republic*, Apr. 20, 2021.

79. Kevin Dobbs, "Despite Biden's Support of Renewables, Enterprise Says Fossil Fuels Vital for Long-Term Energy Needs," *Natural Gas Intelligence*, Feb. 3, 2021, https://www.naturalgasintel.com/despite-bidens-support-of-renewables-enterprise-says-fossil-fuels-vital-for-long-term-energy-needs.

80. European Commission, "What Is the Green New Deal?" https://ec.europa.eu/commission/presscorner/api/files/attachment/859152/What_is_the_European_Green_Deal_en.pdf.pdf; European Commission, "The European Green New Deal," Dec. 11, 2019, https://ec.europa.eu/info/sites/info/files/european-green-deal-communication_en.pdf.

81. Green New Deal for Europe, *A Blueprint for Europe's Just Transition*.

82. IPCC, Fifth Assessment Report, Topic 4, Adaptation and Mitigation, https://ar5-syr.ipcc.ch/topic_adaptation.php; IPCC, Special Report, Headline Statements, https://www.ipcc.ch/sr15/resources/headline-statements.

83. David Roberts, "No Country on Earth Is Taking the 2 Degree Climate Target Seriously," *Vox*, Apr. 29, 2017, https://www.vox.com/2016/10/4/13118594/2-degrees-no-more-fossil-fuels; Oil Change International, *The Sky's Limit: Why the Paris Climate Goals Require a Managed Decline of Fossil Fuel Production* (Sept. 2016), http://priceofoil.org/content/uploads/2016/09/OCI_the_skys_limit_2016_FINAL_2.pdf.

84. Jason Hickel et al., "Urgent Need for Post-Growth Climate Mitigation Scenarios," *Nature Energy* 6, no. 8 (August 2021): 766–68, https://doi.org/10.1038/s41560-021-00884-9.

CHAPTER 3: WHAT CAN I DO AS AN INDIVIDUAL?

1. Seth Wynes and Kimberly A. Nicholas, "The Climate Mitigation Gap: Education and Government Recommendations Miss the Most Effective Individual Actions," *Environmental Research Letters* 12 (2017); David Roberts, "The Best Way to Reduce Your Personal Carbon Emissions: Don't Be Rich," *Vox*, Oct. 15, 2018, https://www.vox.com/energy-and-environment/2017/7/14/15963544/climate-change-individual-choices.

2. Zeke Hausfather, "Factcheck: How Electric Vehicles Help to Tackle Climate Change," *Carbon Brief*, May 13, 2019, https://www.carbonbrief.org/factcheck-how-electric-vehicles-help-to-tackle-climate-change.

3. See Business & Human Rights Resource Centre, "Human Rights in the Mineral Supply Chains of Electric Vehicles," n.d., https://dispatches.business-humanrights.org/human-rights-in-the-mineral-supply-chains-of-electric-vehicles/index.html.

4. See Ray Galvin, "Who Co-opted Our Energy Efficiency Gains? A Sociology of Macro-Level Rebound Effects and U.S. Car Makers," *Energy Policy* 142 (July 2020).

5. Riofrancos, "What Green Costs."

6. Mikhail Chester, Arpad Horvath, and Samer Madanat, "Parking Infrastructure: Energy, Emissions, and Automobile Lifecycle Environmental Accounting," *Environmental Research Letters* 5, no. 3 (July 2010), https://iopscience.iop.org/article/10.1088/1748-9326/5/3/034001.

7. Jacob Poushter, "Car, Bike, or Motorcycle? Depends on Where You Live," Pew Research Center, Apr. 16, 2015, https://www.pewresearch.org/fact-tank/2015/04/16/car-bike-or-motorcycle-depends-on-where-you-live.

8. US Energy Information Administration, "Annual Passenger Travel Tends to Increase with Income," *Today in Energy*, May 11, 2016, https://www.eia.gov/today inenergy/detail.php?id=26192.

9. Enrique Peñalosa, "Why Buses Represent Democracy in Action," TED Talk, https://www.youtube.com/watch?time_continue=58&v=j3YjeARuilI&feature=emb_logo. Also available at http://www.morethangreen.es/enrique-penalosa-y-los-autobuses-en-bogota.

10. Nelson D. Chan and Susan A. Shaheen, "Ridesharing in America: Past, Present, and Future," *Transport Reviews* 32, no. 1 (Jan. 2012): 93–112, gives a history of actual ridesharing, something very different from what Uber and Lyft offer.

The Associated Press stylebook has repeatedly urged that Uber and Lyft be termed "ride-hailing" or "ride-booking" rather than ride-sharing services. See AP Style-Chat, July 17, 2018, https://twitter.com/apstylebook/status/1019291603276652544?lang=en. Shaheen suggests "ride-sourcing" to emphasize the role of the online platform that "sources" the driver from a pool. See Susan Shaheen, Nelson Chan, and Lisa Rayle, "Ridesourcing's Impact and Role in Urban Transportation," *ACCESS* 51 (Spring 2017), https://www.accessmagazine.org/spring-2017/ridesourcings-impact-and-role-in-urban-transportation.

11. Steven Hill, "Ridesharing Versus Public Transit," *American Prospect*, Mar. 27, 2018, https://prospect.org/infrastructure/ridesharing-versus-public-transit; Robert Greenwald, "Walmart: The High Cost of a Low Price" (2005), https://www.bravenewfilms.org/walmartmovie.

12. Miranda Katz, "Why Are New York City Taxi Drivers Killing Themselves?" *Wired*, Mar. 28, 2018, https://www.wired.com/story/why-are-new-york-taxi-drivers-committing-suicide; Michael Goldstein, "Dislocation and its Discontents: Ride-Sharing's Impact on the Taxi Industry," *Forbes*, June 8, 2018.

13. Hill, "Ridesharing Versus Public Transit."

14. Manya Kidambi, "Popularity of Brief Uber, Lyft Rides on Campus Raises Environmental Concerns," *Daily Bruin*, Jan. 29, 2019, https://dailybruin.com/2019/01/29/popularity-of-brief-uber-lyft-rides-on-campus-raises-environmental-concerns; Bruce Schaller, "The New Automobility: Lyft, Uber, and the Future of American Cities," Schaller Consulting, July 25, 2018, http://www.schallerconsult.com/rideservices/automobility.htm; Andrew J. Hawkins, "Uber and Lyft Are the 'Biggest Contributors' to San Francisco's Traffic Congestion, Study Shows," *The Verge*, May 8, 2019; Bruce Schaller, "In a Reversal, 'Car-Rich' Households Are Growing," *CityLab*, Jan. 7, 2019, https://www.citylab.com/perspective/2019/01/uber-lyft-make-traffic-worse-more-people-own-cars-transit/579481.

15. Hill, "Ridesharing Versus Public Transit"; Angie Schmitt, "Study: Uber and Lyft Cause U.S. Transit Decline," *StreetsblogUSA*, Jan. 22, 2019, https://usa.streetsblog.org/2019/01/22/study-uber-and-lyft-are-responsible-for-u-s-transit-decline; Michael Graehler Jr., Richard Alexander Mucci, and Gregory D. Erhardt, "Understanding Recent Transit Ridership Decline in Major U.S. Cities: Service Cuts, or Emerging Modes?" presented at the 98th Annual Meeting, Transportation Research Board, 2019, https://www.researchgate.net/publication/330599129_Understanding_the_Recent_Transit_Ridership_Decline_in_Major_US_Cities_Service_Cuts_or_Emerging_Modes.

16. Schaller, "In a Reversal, 'Car-Rich' Households Are Growing."

17. Stephan Gösling and Paul Upham, "Introduction: Aviation and Climate Change in Context," in Gösling and Upham, eds., *Climate Change and Aviation: Issues, Challenges, and Solutions* (London: Earthscan, 2009), 4–5; Arthur Sullivan, "To Fly or Not to Fly? The Environmental Cost of Air Travel," *DW*, Jan. 4, 2020, https://www.dw.com/en/to-fly-or-not-to-fly-the-environmental-cost-of-air-travel/a-42090155; Seth Wynes and Kimberly A. Nicholas, "The Climate Mitigation Gap: Education and Government Recommendations Miss the Most Effective Individual Actions," *Environmental Research Letters* 12 (2017).

18. Sullivan, "To Fly or Not to Fly?" *DW*.

19. Duygu Yengin and Tracey Dodd, "Flight Shame Won't Fix Airline Emissions. We Need a Smarter Solution," *The Conversation*, Jan. 14, 2020.

20. Umair Irfan, "Air Travel Is a Huge Contributor to Climate Change. A New Global Movement Wants You to Be Ashamed to Fly," *Vox*, Nov. 30, 2019, https://www.vox.com/the-highlight/2019/7/25/8881364/greta-thunberg-climate-change-flying-airline.

21. US Environmental Protection Agency, "Sources of Greenhouse Gas Emissions," https://www.epa.gov/ghgemissions/sources-greenhouse-gas-emissions; US Environmental Protection Agency, "Global Greenhouse Gas Emissions Data," https://www.epa.gov/ghgemissions/global-greenhouse-gas-emissions-data; for an IPCC special report on climate change, desertification, land degradation, sustainable land management, food security, and greenhouse gas fluxes in terrestrial ecosystems, see P. R. Shukla, J. Skea, E. Calvo Buendia, V. Masson-Delmotte, H.-O. Pörtner, D. C. Roberts, P. Zhai, R. Slade, S. Connors, R. van Diemen, M. Ferrat, E. Haughey, S. Luz, S. Neogi, M. Pathak, J. Petzold, J. Portugal Pereira, P. Vyas, E. Huntley, K. Kissick, M. Belkacemi, J. Malley, eds., "Summary for Policymakers," in *Climate Change and Land: An IPCC Special Report on Climate Change, Desertification, Land Degradation, Sustainable Land Management, Food Security, and Greenhouse Gas Fluxes in Terrestrial Ecosystems* (New York: IPCC, 2019), https://www.ipcc.ch/srccl, 8.

22. Hannah Ritchie and Max Roser, "Environmental Impacts of Food Production," Our World in Data, Jan. 2020, https://ourworldindata.org/environmental-impacts-of-food.

23. Ritchie and Roser, "Environmental Impacts of Food Production"; Walter Willett et al., "Food in the Anthropocene: The EAT-Lancet Commission on Healthy Diets and Sustainable Food Systems," The *Lancet* Commissions 393, no. 10170 (Feb. 2019): 447-92, 471; Simon Worrall, "Eating a Burger or Driving a Car: Which Harms the Planet More?" *National Geographic*, Mar. 11, 2015; UN Food and Agriculture Organization, *Livestock's Long Shadow: Environmental Issues and Options* (Rome: FAO, 2006), xxi.

24. Dave Merrill and Lauren Leatherby, "Here's How America Uses Its Land," *Bloomberg*, July 31, 2018, https://www.bloomberg.com/graphics/2018-us-land-use.

25. Frances Moore Lappé, *Diet for a Small Planet*, 20th-ann. ed. (New York: Ballantine Books, 1991), 9–10; Tony Weis, *The Ecological Hoofprint: The Global Burden of Industrial Livestock* (London: Zed Books, 2013), 4.

26. Willett et al., "Food in the Anthropocene," 470, 472; Ritchie and Roser, "Environmental Impacts of Food Production."

27. Oxfam, "The Hunger Virus: How Covid-19 Is Fuelling Hunger in a Hungry World," media briefing, July 2020, https://www.oxfam.org/en/research/hunger-virus-how-covid-19-fuelling-hunger-hungry-world; Shukla et al., "Summary for Policymakers," *Climate Change and Land*.

28. Timothy A. Wise, *Eating Tomorrow: Agribusiness, Family Farmers, and the Battle for the Future of Food* (New York: New Press, 2019).

29. Lappé, *Diet for a Small Planet*, 11.

30. Lappé, *Diet for a Small Planet*, xvii.

31. Hannah Ritchie and Max Roser, "Meat and Dairy Production," Our World in Data, Aug. 2017 (rev. Nov. 2019), https://ourworldindata.org/meat-production.

32. Abram Lustgarten, "Where Will Everyone Go?" *ProPublica* and *New York Times Magazine*, July 23, 2020, https://features.propublica.org/climate-migration/model-how-climate-refugees-move-across-continents.

33. See Todd Miller, *Storming the Wall: Climate Change, Migration, and Homeland Security* (San Francisco: City Lights Books, 2017), 28.

34. Masson-Delmotte et al., "Summary for Policymakers," in *Global Warming of 1.5°C*, https://www.ipcc.ch/sr15/chapter/spm.

35. UN Food and Agriculture Organization, "The 10 Elements of Agroecology: Guiding the Transition to Sustainable Food and Agricultural Systems," 2018, http://www.fao.org/documents/card/en/c/I9037EN.

36. Via Campesina, "La Via Campesina in Action," *La Via Campesina*, Oct. 7, 2018, https://viacampesina.org/en/publication-la-via-campesina-in-action-for-climate-justice-radical-realism-for-climate-justice, 9.

37. GRAIN, "Hungry for Land: Small Farmers Feed the World with Less Than a Quarter of the World's Farmland," May 28, 2014, https://www.grain.org/article/entries/4929-hungry-for-land-small-farmers-feed-the-world-with-less-than-a-quarter-of-all-farmland#sdfootnote37sym. See also GROUP, "Who Will Feed Us? The Peasant Food Web vs the Industrial Food Chain," 3rd ed., 2017, http://www.etcgroup.org/sites/www.etcgroup.org/files/files/etc-whowillfeedus-english-webshare.pdf.

38. Forum for Food Sovereignty, "Declaration of Nyéléni," Feb. 2007, https://nyeleni.org/DOWNLOADS/Nyelni_EN.pdf.

39. CUNY Urban Food Policy Institute, "Putting Food on the Green New Deal Menu: A 7-Point Plan," Sept. 19, 2019, https://www.cunyurbanfoodpolicy.org/news/2019/9/15/putting-food-on-the-green-new-deal-menu-a-7-point-plan.

40. #keepitintheground, "Over 400 Organizations Call on World Leaders: End New Fossil Fuel Development," http://keepitintheground.org; Christophe McGlade and Paul Elkins, "The Geographical Distribution of Fossil Fuels Unused When Limiting Global Warming to 2°C," *Nature* 517 (2015): 187–90; Dan Tong, Quaing Zhang, Yixuan Zheng, et al., "Committed Emissions from Existing Energy Infrastructure Jeopardize 1.5 °C Climate Target," *Nature* 572 (2019): 373–77.

41. "Our Russian 'Pipeline' and Its Ugly Toll," editorial, *Boston Globe*, Feb. 13, 2018.

42. Stefan Andreasson, "Fossil Fuel Divestment Will Increase Carbon Emissions, Not Lower Them—Here's Why," *The Conversation*, Nov. 26, 2019, http://theconversation.com/fossil-fuel-divestment-will-increase-carbon-emissions-not-lower-them-heres-why-126392; C. J. Polychroniou, "Are Fossil Fuel Divestment Campaigns Working? A Conversation with Economist Robert Pollin," *Global Policy Journal*, May 29, 2018, https://www.globalpolicyjournal.com/blog/29/05/2018/are-fossil-fuel-divestment-campaigns-working-conversation-economist-robert-pollin; Charles Komanoff, "The Divestment Diversion," Carbon Tax Center, July 10, 2019, https://www.carbontax.org/blog/2019/07/10/the-divestment-diversion; John Mulliken, "Big Oil Gets Clean and the World Stays Dirty," *Boston Globe*, June 10, 2021.

43. GoFossilFree, "What Is Fossil Fuel Divestment?" https://gofossilfree.org/divestment/what-is-fossil-fuel-divestment.

44. Masson-Delmotte et al., "Summary for Policymakers," in *Global Warming of 1.5°C*, https://www.ipcc.ch/sr15/chapter/spm.

45. Juliet B. Schor and Craig J. Thompson, "Introduction: Practicing Plenitude," in *Sustainable Lifestyles and the Quest for Plenitude: Case Studies in the New Economy*, ed. Craig J. Thompson and Juliet B. Schor (New Haven, CT: Yale University Press, 2014), 10.

46. See, for example, Larry Elliott, "Millennials May Be First to Earn Less Than Previous Generation: Study," *Guardian*, July 18, 2016.

CHAPTER 4: SOCIAL, RACIAL, AND ECONOMIC JUSTICE

1. Martínez-Alier published early works under the Spanish version of his first name, Juan, but now uses the Catalan version, Joan.

2. International Labor Organization, Regions and Countries, Arab States, Areas of Work, Labor Migration, https://www.ilo.org/beirut/areasofwork/labour-migration/lang--en/index.htm.

3. NAACP, "Criminal Justice Fact Sheet," https://www.naacp.org/criminal-justice-fact-sheet; Kimberley Amadeo, "Racial Wealth Gap in the United States," *The Balance*, June 25, 2019, https://www.thebalance.com/racial-wealth-gap-in-united-states-4169678.

4. See GermanWatch, "Global Climate Risk Index 2020," https://www.germanwatch.org/en/17307.

5. See Todd Miller, *Storming the Wall: Climate Change, Migration, and Homeland Security* (San Francisco: City Lights Books, 2017); Nina von Uexkull et al., "Civil Conflict Sensitivity to Growing-Season Drought," *Proceedings of the National Academy of Sciences* 113, no. 44 (Nov. 1, 2016): 12391–96, https://doi.org/10.1073/pnas.1607542113.

6. Naomi Klein, *The Shock Doctrine: The Rise of Disaster Capitalism* (New York: Picador/Henry Holt, 2007), 6.

7. World Meteorological Organization, *State of the Climate in Africa 2019* (Geneva: WMO, 2020), 3, https://library.wmo.int/index.php?lvl=notice_display&id=21778#.X9Iu1bNOk2y.

8. Jonathan Woetzel et al., *Climate Risk and Response in Asia* (McKinsey Global Institute, Nov. 24, 2020), https://www.mckinsey.com/business-functions/sustainability/our-insights/climate-risk-and-response-in-asia#.

9. Christopher B. Field et al., eds., "Central and South America," in *Climate Change 2014: Impacts, Adaptation, and Vulnerability*, Working Group II Contribution to the Fifth Assessment Report of the Intergovernmental Panel on Climate Change (Cambridge: Cambridge University Press/IPCC, 2014), chapter 27, https://www.ipcc.ch/report/ar5/wg2.

10. A. Jay, D. R. Reidmiller, C. W. Avery, D. Barrie, B. J. DeAngelo, A. Dave, M. Dzaugis, M. Kolian, K. L. M. Lewis, K. Reeves, and D. Winner, "Overview," in *Impacts, Risks, and Adaptation in the United States: Fourth National Climate Assessment, Volume II*, ed. D. R. Reidmiller, C. W. Avery, D. R. Easterling, K. E. Kunkel,

K. L. M. Lewis, T. K. Maycock, and B. C. Stewart (Washington, DC: US Global Change Research Program, 2018), 33–71.

11. Rob Wallace, *Big Farms Make Big Flu: Dispatches on Agriculture, Infectious Disease, and the Nature of Science* (New York: Monthly Review Press, 2016), 12, 11.

12. Rob Wallace, *Dead Epidemiologists: On the Origins of COVID-19* (New York: Monthly Review Press, 2020), 55.

13. Wallace, *Big Farms Make Big Flu*, 48.

14. Wallace, *Big Farms Make Big Flu*, 29.

15. Richard A. Oppel Jr., Robert Gebeloff, K. K. Rebecca Lai, Will Wright, and Mitch Smith, "The Fullest Look Yet at the Racial Inequality of the Coronavirus," *New York Times*, July 5, 2020.

16. Vanessa Colón Almenas, Víctor Rodríguez Velázquez, McNelly Torres, and Coral Murphy, "How COVID-19 Hit Puerto Rican New Yorkers Hard in the Bronx and Beyond," *The City*, June 28, 2020, https://www.thecity.nyc/coronavirus/2020/6/28/21306177/covid-19-hit-puerto-rican-new-yorkers-in-bronx.

17. Deanna Pan and John Hancock, "In Suffolk County Black and Latino Residents Face Stark Disparities in Vaccine Access," *Boston Globe*, Jan. 23, 2021; Julia Belluz, "Poorer Countries Might Not Get Vaccinated Until 2023," *Vox*, Apr. 29, 2021, https://www.vox.com/2021/4/28/22405279/covid-19-vaccine-india-covax.

18. See Hannah Ritchie and Max Roser, "CO2 and Greenhouse Gas Emissions," Our World in Data, updated Dec. 2019, https://ourworldindata.org/co2-and-other-greenhouse-gas-emissions.

19. Worldometers, "Luxembourg CO2 Emissions," https://www.worldometers.info/co2-emissions/luxembourg-co2-emissions.

20. Worldometers, "CO2 Emissions," https://www.worldometers.info/co2-emissions.

21. Worldometers, "CO2 Emissions."

22. Lisa Moore, "Greenhouse Gases: How Long Will They Last?" Environmental Defense Fund, Feb. 26, 2008, http://blogs.edf.org/climate411/2008/02/26/ghg_lifetimes.

23. Ritchie and Roser, "CO2 and Greenhouse Gas Emissions."

24. Jason Hickel, "Quantifying National Responsibility for Climate Breakdown: An Equality-Based Attribution Approach for Carbon Dioxide Emissions in Excess of the Planetary Boundary," *Lancet*, Sept. 2020, https://www.thelancet.com/journals/lanplh/article/PIIS2542-5196(20)30196-0/fulltext.

25. Zeke Hausfather, "Mapped: The World's Largest CO2 Importers and Exporters," *Carbon Brief*, July 5, 2017, https://www.carbonbrief.org/mapped-worlds-largest-co2-importers-exporters; Daniel Moran, Ali Hasanbeigi, and Cecilia Springer, "The Carbon Loophole in Climate Policy: Quantifying the Embodied Carbon in Traded Products," *KGM & Associates and Global Efficiency Intelligence* (Aug. 2018): 7, https://buyclean.org/media/2016/12/The-Carbon-Loophole-in-Climate-Policy-Final.pdf.

26. Hannah Ritchie, "How Do CO2 Emissions Compare When We Adjust for Trade?" Our World in Data, Oct. 7, 2019, https://ourworldindata.org/consumption-based-co2; Hausfather, "Mapped."

27. Lucas Chancel and Thomas Piketty, "Carbon and Inequality: From Kyoto to Paris," Paris School of Economics, Nov. 2015, https://www.researchgate.net/publication/285206440_Carbon_and_inequality_From_Kyoto_to_Paris.

28. Chancel and Piketty, "Carbon and Inequality."

29. See Tom Phillips, "China to Move Millions of People from Homes in Anti-Poverty Drive," *Guardian*, Jan. 7, 2018.

30. Dominik Wiedenhofer, Dabo Guan, Zhu Liu, Jing Meng, Ning Zhang, and Yi-MingWei, "Unequal Household Carbon Footprints in China," *Nature Climate Change* 3165 (Dec. 19, 2016), https://scholar.harvard.edu/files/zhu/files/nclimate3165_1.pdf; Kaihui Song et al., "Scale, Distribution and Variations of Global Greenhouse Gas Emissions Driven by U.S. Households," *Environment International* 133 (Dec. 1, 2019): 105137, https://doi.org/10.1016/j.envint.2019.105137; Benjamin Goldstein, Dimitrios Goundaridis, and Joshua P. Newell, "The Carbon Footprint of Energy Use in the United States," *Proceedings of the National Academy of Sciences of the United States*, July 20, 2020, https://www.pnas.org/content/early/2020/07/14/1922205117.

31. Niccolò Manych, Jan Christoph Steckel, and Michael Jakob, "How Finance from Rich Nations Could Drive 40% of New Coal Plant Emissions," *Carbon Brief*, Apr. 14, 2021, https://www.carbonbrief.org/guest-post-how-finance-from-rich-nations-could-drive-40-of-new-coal-plant-emissions; Oil Change International, "Talk Is Cheap: How G20 Governments Are Financing Climate Disaster," July 2017, http://priceofoil.org/2017/07/05/g20-financing-climate-disaster; Tim Quinson and Mathieu BenHamou, "Banks Always Backed Fossil Fuel over Green Projects—Until This Year," *Bloomberg*, May 19, 2021.

32. Paul Griffin, "The Carbon Majors Database," Corporate Citizenship Briefing, July 19, 2017, https://ccbriefing.corporate-citizenship.com/2017/07/19/carbon-majors-database-100-companies-responsible-71-global-emissions.

33. Climate Accountability Institute, "Carbon Majors," Oct. 8, 2019, https://climateaccountability.org/carbonmajors.html.

34. Matthew Taylor and Jonathan Watts, "Revealed: The 20 Firms Behind a Third of All Carbon Emissions," *Guardian*, Oct. 9, 2019.

35. Kelly Levin, "World's Carbon Budget to Be Spent in Three Decades," World Resources Institute, Sept. 27, 2013, https://www.wri.org/blog/2013/09/world-s-carbon-budget-be-spent-three-decades; Roz Pidcock, "Carbon Briefing: Making Sense of the IPCC's New Carbon Budget," *CarbonBrief*, Oct. 23, 2013, https://www.carbonbrief.org/carbon-briefing-making-sense-of-the-ipccs-new-carbon-budget; Zeke Hausfather, "Analysis: Why Children Today Must Emit Eight Times Less CO2 Than Their Grandparents," *CarbonBrief*, Apr. 10, 2019, https://www.carbonbrief.org/analysis-why-children-must-emit-eight-times-less-co2-than-their-grandparents; Axel Dalman, "Carbon Budgets: Where Are We Now?" Carbontracker.org, May 11, 2020, https://carbontracker.org/carbon-budgets-where-are-we-now; Daniel W. O'Neill, Andrew L. Fanning, William F. Lamb, and Julia K. Steinberger, "A Good Life for All Within Planetary Boundaries," *Nature Sustainability* 1 (2018): 88–95.

36. Christian Holz, Sivan Kartha, and Tom Athanasiou, "Fairly Sharing 1.5: National Fair Shares of a 1.5°C-Compliant Global Mitigation Effort," *International Environmental Agreements: Politics, Law and Economics* 18, no. 1 (2018): 117–34; Greg Muttitt and Sivan Kartha, "Equity, Climate Justice, and Fossil Fuel Extrac-

tion: Principles for a Managed Phase Out," *Climate Policy* (2020); Hickel, "Quantifying National Responsibility for Climate Breakdown"; Tom Athanasiou, "Over 50,000 People and 195 Global Groups Demand Biden Commit the US to Do Its 'Fair Share' on Climate," *US Climate Fair Share*, Feb. 21, 2021, https://usfairshare.org/blog/over-50000-people-195-global-groups-demand-biden-commit-the-u-s-to-do-its-fair-share-on-climate.

37. Marie Gottschalk, *The Shadow Welfare State: Labor, Business, and the Politics of Health Care in the United States* (Ithaca, NY: ILR Press, 2000); Matthew Walters and Lawrence Mishel, "How Unions Help All Workers," Economic Policy Institute, Aug. 26, 2003, https://www.epi.org/publication/briefingpapers_bp143.

38. See Lawrence B. Glickman, *A Living Wage: American Workers and the Making of Consumer Society* (Ithaca, NY: Cornell University Press, 1999); Dana Frank, *Buy American! The Untold Story of Economic Nationalism* (Boston: Beacon Press, 2000); Lizabeth Cohen, *A Consumers' Republic: The Politics of Mass Consumption in Postwar America* (New York: Alfred A. Knopf, 2003); Matthew T. Huber, *Lifeblood: Oil, Freedom, and the Forces of Capital* (Minneapolis: University of Minnesota Press, 2013).

39. See Dorceta E. Taylor, *Toxic Communities: Environmental Racism, Toxic Communities, and Residential Mobility* (New York: New York University Press, 2014); Bruce Mayer, *Blue-Green Coalitions: Fighting for Safe Workplaces and Healthy Communities* (Ithaca, NY: ILR Press, 2008); Chad Montrie, *Making a Living: Work and Environment in the United States* (Chapel Hill: University of North Carolina Press, 2008); and Joan Martínez-Alier, *Environmentalism of the Poor: A Study of Ecological Conflicts and Valuation* (Cheltenham, UK: Edward Elgar, 2003).

40. See Jake Rosenfeld, Patrick Denice, and Jennifer Laird, "Union Decline Lowers Wages of Nonunion Workers," Economic Policy Institute, Aug. 30, 2016, https://www.epi.org/publication/union-decline-lowers-wages-of-nonunion-workers-the-overlooked-reason-why-wages-are-stuck-and-inequality-is-growing.

41. See Chad Montrie, *A People's History of Environmentalism in the United States* (London: Continuum International Publishing Group, 2011).

42. See AFL-CIO, Executive Council Statement, "Kyoto Protocol," Jan. 30, 1998, https://aflcio.org/about/leadership/statements/kyoto-protocol; Sean Sweeney, "Climate Change and the Great Inaction: New Trade Union Perspectives," Trade Unions for Energy Democracy, Sept. 2014, http://unionsforenergydemocracy.org/wp-content/uploads/2014/09/TUED-working-paper-2.pdf.

43. Sean Sweeney, "Pandering to the Predator: Labor and Energy Under Trump," *New Labor Forum* (Feb. 2017), https://newlaborforum.cuny.edu/2017/02/03/pandering-to-the-predator-labor-and-energy-under-trump; Jane McAlevey, "Organizing to Win a Green New Deal," *Jacobin*, Mar. 26, 2019, https://www.jacobinmag.com/2019/03/green-new-deal-union-organizing-jobs.

44. BlueGreen Alliance, "Solidarity for Climate Action," http://www.bluegreenalliance.org/wp-content/uploads/2019/07/Solidarity-for-Climate-Action-vFINAL.pdf.

45. Sweeney, "Climate Change and the Great Inaction," 16; Unions Against Fracking, "We Call for a Global Moratorium," http://unionsagainstfracking.org/the-statement; Alexander C. Kaufman, "Green New Deal Picks Up Two Major

Union Endorsements," *Huffington Post*, June 7, 2019, https://www.huffpost.com/entry/green-new-deal-unions_n_5cfa9b5ee4b0aab91c0589c7.

46. Labor Network for Sustainability, "Mission and Principles," http://www.labor4sustainability.org/wp-content/uploads/2017/03/LNS-Mission-and-Principles.pdf.

47. Cecil E. Roberts (International President, United Mine Workers of America) and Lonnie R. Stephenson (International President, International Brotherhood of Electrical Workers) to Sen. Edward Markey and Rep. Alexandria Ocasio-Cortez, Mar. 8, 2019, http://www.ibew.org/Portals/22/IBEW%20Letters/2019/Markey.Ocasio-Cortez%20Letter.Climate.pdf?ver=2019-03-11-111951-970.

48. Friends of the Earth, "Environmentalists Call on AFL-CIO to Support Green New Deal," press release, Mar. 13, 2019, https://foe.org/news/environmentalists-call-afl-cio-support-green-new-deal.

49. This section relies heavily on Labor Network for Sustainability, "Just Transition: Just What Is It?" https://www.labor4sustainability.org/uncategorized/just-transition-just-what-is-it. For the GI Bill, see US Department of Veterans Affairs, Education and Training, History and Timeline, https://www.benefits.va.gov/gibill/history.asp.

50. Labor Network for Sustainability, "Just Transition."

51. Trumka was referring to a set of programs called Trade Adjustment Assistance, which are supposed to compensate workers whose jobs are lost due to imports as protective tariffs and laws have been dismantled. Richard Trumka, "Trumka to Utility Workers: The Only Way to Go Is Up," speech, Apr. 13, 2016, https://aflcio.org/speeches/trumka-utility-workers-only-way-go; Labor Network for Sustainability, "Just Transition"; Brad Markell, personal communication, Aug. 7, 2019.

52. Ashley Dawson, *People's Power: Reclaiming the Energy Commons* (New York: OR Books, 2020), 13.

53. Eora Global Supply Chain Database, "Carbon Footprint of Nations," https://worldmrio.com/footprints/carbon.

54. Worldometers, "Norway CO2 Emissions," https://www.worldometers.info/co2-emissions/norway-co2-emissions; Lars Petter Teigen, "Norway's Green Delusions," *Foreign Policy*, Sept. 19, 2018, https://foreignpolicy.com/2018/09/19/norways-green-delusions-oil-gas-drilling.

55. Sweeney, "Climate Change and the Great Inaction," 14.

56. Sweeney, "Climate Change and the Great Inaction," 13; Sean Sweeney, *Resist, Reclaim, Restructure: Unions and the Struggle for Energy Democracy* (Rosa Luxemburg Stiftung, 2013), https://www.rosalux.de/en/publication/id/8419/resist-reclaim-restructure-1, 38.

57. Kate Aronoff, Alyssa Battistoni, Daniel Aldana Cohen, and Thea Riofrancos, *A Planet to Win: Why We Need a Green New Deal* (London: Verso, 2019), 108; Sweeney, *Resist, Reclaim, Restructure*, 44–45.

58. TransNational Institute, *Towards Energy Democracy: Discussions and Outcomes from an International Workshop* (Amsterdam: TransNational Institute, May 2016), https://www.tni.org/en/publication/towards-energy-democracy.

59. UN Food and Agriculture Organization, "Food Security and Nutrition Around the World in 2020," http://www.fao.org/3/ca9692en/online/ca9692en.html#chapter-executive_summary.

60. Barbara J. Fields and Karen E. Fields, "Did the Color of His Skin Kill Philando Castile?" *Jacobin*, July 13, 2016, https://www.jacobinmag.com/2016/07/racecraft-barbara-karen-fields-philando-castile.

61. Aronoff, et al., *A Planet to Win*, 27.

CHAPTER 5: BROADENING THE LENS

1. Brady E. Hamilton, Joyce A. Martin, Michelle J. K. Osterman, and Lauren M. Rossen, *Vital Statistics Rapid Release: Births: Provisional Data for 2019*, report 007 (Washington, DC: US Department of Health and Human Services, Division of Vital Statistics, National Center for Health Statistics, May 2019), https://www.cdc.gov/nchs/data/vsrr/vsrr-007-508.pdf; Bill Chapell, "U.S. Birthrates Fell to a 32-Year Low in 2013; CDC Says Birthrate Is in Record Slump," NPR, May 15, 2019.

2. Eurostat, "Fertility Statistics," https://ec.europa.eu/eurostat/statistics-explained/index.php/Fertility_statistics; World Bank, "Fertility Rate, Total (Births per Woman), UN Population Division, World Population Prospects: 2019 Revision, https://data.worldbank.org/indicator/SP.DYN.TFRT.IN.

3. US Census, "U.S. and World Population Clock," https://www.census.gov/popclock.

4. Links at Worldometer, "Population: World," https://www.worldometers.info/population/world; "The Lancet: World Population Likely to Shrink After Mid-Century, Forecasting Major Shifts in Global Population and Economic Power," *ScienceDaily*, July 15, 2020, www.sciencedaily.com/releases/2020/07/200715150444.htm.

5. Eunice Meuni, "Changing the Narrative on Fertility Decline in Africa," *Wilson Center NewSecurityBeat*, Apr. 20, 2016, https://www.newsecuritybeat.org/2016/04/changing-narrative-fertility-decline-africa; United Nations, *World Fertility Patterns 2015*, data booklet, https://www.un.org/en/development/desa/population/publications/pdf/fertility/world-fertility-patterns-2015.pdf.

6. Andreas Malm, *Fossil Capital: The Rise of Steam Power and the Roots of Global Warming* (London: Verso, 2014), 268–69.

7. Gaby del Valle, "When Environmentalism Meets Xenophobia," *Nation*, Nov. 8, 2018, https://www.thenation.com/article/environment-climate-eugenics-immigration.

8. Bill McKibben, "Does It Make Sense for Environmentalists to Want to Limit Immigration?" *Grist*, Mar. 2, 2004, https://grist.org/article/mckibben-immigration.

9. Thanu Yatupikiyage, "If You Care About Climate Change, You Should Care About Anti-Immigrant Policy," 350.org, June 27, 2018, https://350.org/climate-change-and-immigration; Hop Hopkins, "How the Sierra Club's History with Immigrant Rights Is Changing Our Future," *Sierra Club*, Nov. 2, 2018, https://www.sierraclub.org/articles/2018/11/how-sierra-club-s-history-immigrant-rights-shaping-our-future; Bill McKibben, "Protesting Immigration Policy, and Why I Decided to Get Arrested," *New Yorker*, Aug. 9, 2019.

10. Jeanne Batalova, Mary Hanna, and Christopher Levesque, "Frequently Requested Statistics on Immigrants and Immigration in the United States," Migration Policy Institute, Feb. 11, 2021, https://www.migrationpolicy.org/article/frequently-requested-statistics-immigrants-and-immigration-united-states-2020.

11. Gretchen Livingston, "Hispanic Women No Longer Account for the Majority of Immigrant Births in the U.S.," Pew Research Center, Aug. 8, 2019, https://www.pewresearch.org/fact-tank/2019/08/08/hispanic-women-no-longer-account-for-the-majority-of-immigrant-births-in-the-u-s.

12. Rubén Berríos, *Growth Without Development: Peru in Comparative Perspective* (Lanham, MD: Lexington Books, 2019).

13. C. Le Quéré, R. B. Jackson, M. W. Jones, et al. "Temporary Reduction in Daily Global CO2 Emissions During the COVID-19 Forced Confinement," *Nature Climate Change* 10 (2020): 647–53.

14. See Robert Pollin, "Degrowth Versus Green New Deal: A Response to Juliet Schor and Andrew Jorgenson," *Review of Radical Political Economics* 5, no. 2 (2019): 330–32; Dean Baker, "Saving the Environment: Is Degrowthing the Answer?" Center for Economic Policy Research, Nov. 24, 2018, https://cepr.net/saving-the-environment-is-degrowthing-the-answer, and "Will Degrowthing Save the Planet?" Center for Economic Policy Research, Dec. 7, 2018, https://cepr.net/will-degrowthing-save-the-planet.

15. US Energy Information Administration, "U.S. Energy-Related CO2 Emissions Increased in 2018 but Will Likely Fall in 2019 and 2020," Jan. 28, 2019, https://www.eia.gov/todayinenergy/detail.php?id=38133.

16. Trevor Houser and Hanna Pitt, "Preliminary U.S. Emissions Estimates for 2019," Rhodium Group, Jan. 7, 2020, https://www.eenews.net/assets/2020/01/07/document_cw_01.pdf; Benjamin Storrow, "Global CO2 Emissions Were Flat in 2019—But Don't Cheer Yet," *Scientific American*, Feb. 12, 2020.

17. See Jacob Assa, "The Financialization of GDP and Its Implications for Macroeconomic Debates," New School for Social Research Department of Economics Working Paper, Oct. 2016, http://www.economicpolicyresearch.org/econ/2016/NSSR_WP_102016.pdf; Jacob Assa, *The Financialization of GDP: Implications for Economic Theory and Policy* (London: Routledge, 2017).

18. Magnus Jiborn, Astrid Kander, Viktoras Kulionis, Hana Nielsen, and Daniel D. Moran, "Decoupling or Delusion? Measuring Emissions Displacement in Foreign Trade," *Global Environmental Change* 49 (Mar. 1, 2018): 27–34, https://doi.org/10.1016/j.gloenvcha.2017.12.006.

19. M. Andersson and I. Lövin, "Decoupling GDP Growth from CO2 Emissions Is Possible," Swedish Foreign Policy News, May 23, 2015, http://www.swemfa.se/2015/05/23/sweden-decoupling-gdp-growth-from-co2-emissions-is-possible; Jiborn et al., "Decoupling or Delusion?"

20. S. Davis and K. Caldeira, "Consumption-Based Accounting of CO2 Emissions," *Proceedings of the National Academy of Sciences* 107, no. 12 (2010): 5687–92; Malm, *Fossil Capital*, 331; G. Peters, J. Minx, C. Weber, and O. Edenhofer, "Growth in Emission Transfers via International Trade from 1990 to 2008," *PNAS* 108, no. 21 (2011): 8903–8; K. Kanemoto, D. Moran, M. Lenzen, and A. Geschke, "International Trade Undermines National Emission Reduction Targets," *Global Environmental Change* 24 (2014): 52–59, 53; Nicolai Baumert, Astrid Kander, Magnus Jiborn, Viktoras Kulionis, Tobias Nielsen, "Global Outsourcing of Carbon Emissions 1995–2009: A Reassessment," *Environmental Science & Policy* 92 (Feb. 2019): 228–36; Gail Cohen, João Tovar Jalles, Prakash Loungani, and Ricardo Marto, "The

Long-Run Decoupling of Emissions and Output: Evidence from the Largest Emitters," IMF Working Paper, 2018, https://www.imf.org/en/Publications/WP/Issues/2018/03/13/The-Long-Run-Decoupling-of-Emissions-and-Output-Evidence-from-the-Largest-Emitters-45688.

21. Marshall David Sahlins, *Stone Age Economics* (1972; London: Routledge, 2011), 11.

22. James C. Scott, *The Moral Economy of the Peasant: Rebellion and Subsistence in Southeast Asia* (New Haven, CT: Yale University Press, 1976), vii, 3.

23. Steven Stoll, *The Great Delusion: A Mad Inventor, Death in the Tropics, and the Utopian Origins of Economic Growth* (New York: Hill and Wang, 2008), 15.

24. W. W. Rostow, *The Stages of Economic Growth: A Non-Communist Manifesto* (Cambridge: Cambridge University Press, 1960); Nick Cullather, *The Hungry World: America's Cold War Battle Against Poverty in Asia* (Cambridge, MA: Harvard University Press, 2010).

25. John Kenneth Galbraith, *The Affluent Society* (1958; Boston: Houghton Mifflin, 1998), 269.

26. See also Richard A. Easterlin, "Does Economic Growth Improve the Human Lot? Some Empirical Evidence," in *Nations and Households in Economic Growth: Essays in Honor of Moses Abramovitz* (New York: Academic Press, 1974), 89–125; and Richard A. Easterlin, "Happiness and Economic Growth: The Evidence" (Nov. 6, 2014), in *Global Handbook of Quality of Life*, USC-INET Research Paper No. 14-03, ed. W. Glatzer, L. Camfield, V. Møller, and M. Rojas (Dordrecht: Springer, 2015), available at SSRN: https://ssrn.com/abstract=2522476.

27. André Gorz, *Ecology as Politics* (Montreal: Black Rose Books, 1980), 13–14.

28. Irene Hernández Velasco, "Hay que cambiar la manera de medir lo que hacemos y quitarle importancia a lo que digan los economistas," BBC News Mundo, Dec. 3, 2020, https://www.bbc.com/mundo/noticias-55086737.

29. John Bellamy Foster, *Marx's Ecology: Materialism and Nature* (New York: Monthly Review Press, 2000) and *Ecology Against Capitalism* (New York: Monthly Review Press, 2002); Nancy Fraser, "Behind Marx's Hidden Abode," *New Left Review* 86 (March–April 2014), https://newleftreview.org/issues/II86/articles/nancy-fraser-behind-marx-s-hidden-abode.

30. Giorgos Kallis, Federicao Demaria, and Giacomo D'Alisa, "Introduction: Degrowth," in *Degrowth: A Vocabulary for a New Era*, ed. D'Alisa, Demaria, and Kallis (New York: Routledge, 2015), 6; Ramachandra Guha and Juan Martínez-Alier, *Varieties of Environmentalism: Essays North and South* (London: Earthscan, 1997), xxi.

31. Kallis, Demaria, and D'Alisa, "Introduction," 11.

32. Our World in Data, "CO2 Emissions per Capita and GDP per Capita, 2018," https://ourworldindata.org/grapher/co2-emissions-vs-gdp; Max Roser, Esteban Ortiz-Ospina, and Hannah Ritchie, "Life Expectancy," Our World in Data, 2013, https://ourworldindata.org/life-expectancy.

33. Marco Deriu, "Conviviality," in D'Alisa, Demaria, and Kallis, *Degrowth*, 77–78.

34. See Artwell Nhemachena, Tapiwa V. Warikandwa, Nkosinothando Mpofu, and Howard Chitimira, "Explosive Economic Minefields in Invisible NeoImperial

Forcefields: An Introduction to Decolonising Economies in Africa," in *Grid-Locked African Economic Sovereignty: Decolonising the Neo-Imperial Socioeconomic and Legal Forcefields in the 21st Century*, ed. Warikandwa, Nhemachena, Mpofu, and Chitimara (Bamenda, Cameroon: Langaa RPCIG, 2019); Jason Hickel, "The Anti-Colonial Politics of Degrowth," *Political Geography* (2021), https://doi.org/10.1016/j.polgeo.2021.102404.

35. See chapters on these movements in D'Alisa, Demaria, and Kallis, *Degrowth*.

36. E. P. Thompson, "Time, Work-Discipline, and Industrial Capitalism," *Past and Present* 38 (Dec. 1967): 56–97; Herbert Gutman, "Work, Culture, and Society in Industrializing America, 1815–1919," *American Historical Review* 78, no. 3 (1973); Marc Edelman, "Bringing the Moral Economy Back In," *American Anthropologist*, New Series, 107, no. 3 (Sept. 2005); Stoll, *Great Delusion*, 166.

37. Herman E. Daly and John B. Cobb Jr., *For the Common Good: Redirecting the Economy Toward Community, the Environment, and a Sustainable Future*, 2nd ed. (Boston: Beacon Press, 1994), appendix.

38. See the most recent report at John F. Helliwell, Richard Layard, Jeffrey Sachs, and Jan-Emmanuel De Neve, eds., *World Happiness Report, 2020* (New York: Sustainable Development Solutions Network, 2020), https://worldhappiness.report/ed/2020.

39. Jason Hickel, "Human Flourishing Doesn't Require Endless GDP Growth," Dec. 23, 2017, https://www.jasonhickel.org/blog/2017/12/23/martin-ravallion-is-wrong-endless-growth-is-not-necessary-for-human-well-being.

40. Jason Hickel, "Stability Without Growth: Keynes in an Age of Climate Breakdown," Center for Economic Policy Research, Dec. 3, 2018, https://cepr.net/stability-without-growth-keynes-in-an-age-of-climate-breakdown.

41. Kate Raworth, "What on Earth Is the Doughnut?" Exploring Doughnut Economics, https://www.kateraworth.com/doughnut.

42. David Roberts, "The Doughnut of Justice: A New Way to Think About Growth," *Grist*, Feb. 22, 2012, https://grist.org/climate-change/the-doughnut-of-justice-a-new-way-to-think-about-growth.

43. Kate Raworth, "Why Degrowth Has Outgrown Its Own Name," *Oxfam Blog*, Dec. 1, 2015, https://www.kateraworth.com/2015/12/01/degrowth/ and Giorgos Kallis, "You're Wrong, Kate. Degrowth Is a Compelling Word," *Oxfam Blog*, Dec. 2, 2105, https://oxfamblogs.org/fp2p/youre-wrong-kate-degrowth-is-a-compelling-word, and Giorgos Kallis, "Yes, We Can Prosper Without Growth: Ten Policy Proposals for the New Left," Common Dreams, Jan. 28, 2015, https://www.commondreams.org/views/2015/01/28/yes-we-can-prosper-without-growth-10-policy-proposals-new-left.

44. Daniel W. O'Neill, Andrew L. Fanning, William F. Lamb, and Julia K. Steinberger, "A Good Life for All Within Planetary Boundaries," *Nature Sustainability* 1 (2018): 88–95; Dan O'Neill, "Is It Possible for Everyone to Live a Good Life Within Our Planet's Limits?" *The Conversation*, Feb. 7, 2018, https://theconversation.com/is-it-possible-for-everyone-to-live-a-good-life-within-our-planets-limits-91421.

45. GNDE, *A Blueprint for Europe's Just Transition*, 2nd ed. (GNDE, Dec. 2019), https://report.gndforeurope.com.

46. GNDE, *A Blueprint for Europe's Just Transition*."

47. See debates between Dean Baker and Jason Hickel (CEPR, 2018) and between Robert Pollin and Juliet Schor and Andrew K. Jorgenson. Schor and Jorgenson, "Is It Too Late for Growth?" and "Response to Bob Pollin," and Pollin, "Degrowth Versus Green New Deal: Response to Juliet Schor and Andrew Jorgenson," *Review of Radical Political Economics* 5, no. 2 (2019): 320–35. See also Kate Aronoff, Alyssa Battistoni, Daniel Aldana Cohen, and Thea Riofrancos, *A Planet to Win: Why We Need a Green New Deal* (London: Verso, 2019), which advocate curbing demand and slowing down the economy without fully embracing degrowth.

48. See, for example, works by Clive Spash at https://www.clivespash.org; Juliet Schor, "Climate, Inequality, and the Need for Reframing Climate Policy," *Review of Radical Political Economics* 47, no. 4 (2015).

49. Nathaniel Rich, *Losing Earth: A Recent History* (New York: Farrar, Straus and Giroux, 2019), 3.

50. Rich, *Losing Earth*, 8, 5.

51. Rich, *Losing Earth*, 171, 180.

52. Rich, *Losing Earth*, 179.

53. Rich, *Losing Earth*, 181. See also Climate Action Tracker: Countries; https://climateactiontracker.org/countries; *Axios*, "The Major Emitters That Are Meeting Their Paris Agreement Pledges," June 1, 2019, https://www.axios.com/paris-agreement-countries-meeting-pledges-1261f497-3ec7-4192-ba21-83ae339762be.html.

54. Robert Paarlberg, *The United States of Excess: Gluttony and the Dark Side of American Exceptionalism* (Oxford: Oxford University Press, 2015).

55. Paarlberg, *The United States of Excess*, 7.

56. See Neil Kaye, "Running Total of Global Fossil Fuel CO2 Emissions Since 1751," May 22, 2019; https://twitter.com/neilrkaye/status/1131140917317050370/photo/1; Hannah Ritchie and Max Roser, "Cumulative CO2 Emissions by World Region," Our World in Data, Dec. 2019, https://ourworldindata.org/grapher/cumulative-co2-emissions-region.

57. Robert McSweeney, "Explainer: Nine 'Tipping Points' That Could Be Triggered by Climate Change," Carbon Brief, Feb. 10, 2020, https://www.carbonbrief.org/explainer-nine-tipping-points-that-could-be-triggered-by-climate-change.

CONCLUSION

1. Masson-Delmotte et al., "Summary for Policymakers," in *Global Warming of 1.5°C*, https://www.ipcc.ch/sr15/chapter/spm.

2. Masson-Delmotte et al., "Summary for Policymakers," in *Global Warming of 1.5°C*.

3. See Jeff Tollefson, "How the Coronavirus Pandemic Slashed Carbon Emissions," *Nature*, May 20, 2020, https://www.nature.com/articles/d41586-020-01497-0; Brandon Specktor, "Global Carbon Emissions Dropped an Unprecedented 17% During the Coronavirus Lockdown—and It Changes Nothing," *Live Science*, May 20, 2020, https://www.livescience.com/carbon-dioxide-reduction-coronavirus-lockdown.html; Shannon Osaka, "Growing Pains: Post-COVID, Should Countries Rethink Their Obsession with Economic Growth?" *Grist*, Aug. 11, 2020, https://grist.org/politics/post-covid-should-countries-rethink-their-obsession-with-economic-growth.

4. Jeff Tollefson, "COVID Curbed Carbon Emissions in 2020—But Not by Much," *Nature* 589, no. 7842 (Jan. 15, 2021): 343, https://doi.org/10.1038/d41586-021-00090-3.

5. Chuck Collins, Omar Ocampo, and Sophia Paslaski, "Billionaire Bonanza 2020: Wealth Windfalls, Tumbling Taxes, and Pandemic Profiteers," Institute for Policy Studies, 2020, https://ips-dc.org/billionaire-bonanza-2020; Americans for Tax Fairness and Institute for Policy Studies, "Billionaire Pandemic Wealth Gains," Apr. 21, 2021, https://ips-dc.org/wp-content/uploads/2021/04/IPS-ATF-Billionaires-13-Month-31-Year-Report-copy.pdf; Oxfam America, "Pandemic Profiteers Exposed," Oxfam media briefing, July 22, 2020, https://www.oxfamamerica.org/explore/research-publications/pandemic-profits-exposed.

6. See Green New Deal for Europe, *A Blueprint for Europe's Just Transition*, 2nd ed., Dec. 2019, https://www.gndforeurope.com; "What Is Degrowth?" https://www.degrowth.info/en/what-is-degrowth; Juan Francisco Salazar, "Buen Vivir: South America's Rethinking of the Future We Want," *The Conversation*, July 23, 2015, https://theconversation.com/buen-vivir-south-americas-rethinking-of-the-future-we-want-44507.